'F[...] [...] [...]rk out how we can best think. [...] [...]he past, the present and the future of our minds. Read – to be amazed.'

– Bettany Hughes

'Hannah Critchlow has written a timely and engaging book about human intelligence and the challenges our brains face in the twenty-first century. It will make you think. It might even change for the better the way you think.'

– Ian Rankin

'[Hannah's] book is a powerful manifesto for the strength of "we" thinking over "me" thinking told in her characteristically erudite, eloquent and entertaining style.'

– Marcus du Sautoy, author of
Thinking Better: The Art of the Shortcut

'Hannah Critchlow's research into collective intelligence, team-work, communication, performance, resilience, ethics etc from a neuroscience perspective is absolutely fascinating. She takes these complex concepts and makes them easily accessible for everyone.'

– Tatjana Marinko, Middle East Director of the London Speaker Bureau

'From startling futuristic speculation to practical exercises in getting in touch with your own routine mental processes, Hannah Critchlow steers us with a sure hand and an unfailingly clear and engaging voice. This is a treasure of a book, exploding some damaging myths and encouraging us to re-imagine the values of relationality and receptivity in our thinking.'

– Baron Rowan Williams

'This is absolutely wonderful, uplifting and soulful. I can't tell you how much we need joined-up thinking – this book and the thing itself. The future of humanity very much depends on how well we embrace these ground-breaking provocative ideas, to focus on the collective 'we' more than the individual "me".'
                                    – Daniel M. Davis, author of *The Secret Body*

Joined-Up Thinking

The Science of Collective Intelligence and its Power to Change Our Lives

Hannah Critchlow

HODDER &
STOUGHTON

First published in Great Britain in 2023 by Hodder & Stoughton
An Hachette UK company

This paperback edition published in 2023

1

A CIP catalogue record for this title is available from the British Library

Paperback ISBN 978 1 529 39843 4
eBook ISBN 978 1 529 39841 0

Typeset in Bembo by Palimpsest Book Production Limited,
Falkirk, Stirlingshire

Printed and bound in Great Britain by Clays Ltd, Elcograf S.p.A.

Hodder & Stoughton policy is to use papers that are natural, renewable
and recyclable products and made from wood grown in sustainable forests.
The logging and manufacturing processes are expected to conform
to the environmental regulations of the country of origin.

Hodder & Stoughton Ltd
Carmelite House
50 Victoria Embankment
London EC4Y 0DZ

www.hodder.co.uk

I am indebted to my young son Max – watching him grow, learn and contribute to the world fills me with pride and wonder.

This Place painted by Alicia Adams

Contents

Prologue

For nearly seventy years from the end of the Second World War, human beings seemed to be getting cleverer. The average IQ score of populations in countries including Britain, France, Japan and Korea showed a roughly three-point rise every decade. This so-called Flynn effect, named after the researcher who pinpointed it, was attributed to the impact of better education and better diet among children growing up in developed societies. Now it seems that the intelligence boom is over. IQ scores hit a peak for those born during the mid-seventies and are now falling across Europe.

In the spirit of full disclosure I should point out that I was a mere twinkle in my parents' eyes in the mid-seventies so I'm part of a cohort born on the downward slide from the supposed peak of human intelligence! But can it really be that we're getting stupider? Are our junk-food diets and too much screen time decimating our cognitive powers? Or is it the tests that are faulty, rather than our brains? Are they perhaps out of date for the world we now live in?

Intelligence researchers, including James Flynn himself, continue to dispute the size and significance of the Flynn effect, and yet this story never quite goes away. In 2018, a decades-long study on Norwegian army conscripts concluded that IQ levels *were* falling; the media seized on it, and the whole conversation started up again. The Flynn effect may not tell us anything definitive about human intelligence, but it tells us plenty about our fixation on and confusion over the 1.5kg of soft grey flesh within our skulls.

When I read another story in the press about falling IQ levels, part of me wonders whether humanity has hit the limit of our brain power. I also ask myself what conventional thinking about cleverness means, if even intelligence researchers can't agree on terms and trends. I go back to questioning the effectiveness of our educational system, and the faith we have in mental agility as a panacea for problems and a marker of individual success. I think about the people I met when I worked in a psychiatric hospital, whose situation inspired me to study the workings of the brain. Above all, I come back to my conviction that our thinking about intelligence needs an update for the twenty-first century.

Many of us are highly invested in the competitive business (and it really is a business) of boosting and proving our intellectual powers. It starts early. Young people are experiencing more anxiety over exam performance than ever before. As part of my work as a science fellow at Magdalene College, University of Cambridge, I've talked with hundreds of high-school students about the use of smart drugs among their peers. They tell me about buying drugs online in an effort to boost their grades. I've attended academic conferences on gene-editing to tweak the cognitive abilities of unborn babies. I've seen US-based companies advertising pre-implantation screening of embryos to assist parents to avoid having 'intellectually disabled' children. I've met people with medical prosthetics implanted in their brains that alter their feelings, mental faculties and how they interact with the world. The emphasis is always on how our individual brain power (and perhaps, at a push, that of our children) can be boosted, in order to be fitter for this competitive game called life.

I've come to believe that the emphasis on individual smartness, as measured by exam performance and IQ tests and promotion at work, is not in our best interests. That it is limiting to most of us and damaging to many. That it isn't even the most effective model for coming up with the innovative solutions to complex problems that we all need. As neuroscience begins to broaden its focus and investigate how our brains work together to communicate and collaborate, there is a whole host of evidence emerging that our society's emphasis on individual intelligence is out of date.

We now know that all brains, even the most agile and successful, are still fundamentally flawed in the way they perceive the world. Every human brain is subject to bias and blind spots, limitations in thinking, emotional contagion and covert influence by other people. We are all, to some extent, less rational and less intelligent than we like to think we are. We are biased against unusual or unconventional cognitive approaches. We get stuck in our own little bubbles, overlooking ideas and people that could disrupt our thinking in useful ways. We don't talk or listen with enough curiosity and patience to actually learn. We pay lip service to the value of collaboration without knowing how to really do effective joined-up thinking, or what it might mean for us if we did. We've got stuck with a view of intelligence that's no longer fit for purpose.

Rather than obsessing over individual exam grades and relying on the undoubted achievements of the people our system labels as exceptionally clever, could we expand our thinking about thinking? Is there a different, more creative, inclusive and efficient way to think about intelligence, to drive innovation and solve problems?

Collective intelligence is precisely the approach we need to overcome our individual brains' limitations and hit new heights. Pooling ideas and gathering different perspectives allows us to tap into the wisdom and experience embodied in a group of people.

Our species' innate drive to share information and seek out new approaches has evolved as a workaround to cope with the gaps in our individual knowledge and perspective. It means that humanity has been practising collective intelligence ever since groups of our ancestors worked together to gather in the harvest; arguably since the first compassionate act that prioritised the wellbeing of the collective over an individual's immediate needs.

Millennia later, digital technology has shifted collective intelligence online, where it has produced the labyrinth of Wikipedia, the global conversation of Twitter and citizen science campaigns to control Ebola in central Africa, and enabled experiments in direct participatory democracy. Idea-sharing and cooperation are literally in our DNA, and are constantly evolving.

This book is the result of my two-year-long deep dive into the cutting-edge neuroscience of humanity's collective intelligence. That journey convinced me that we're at a pivotal moment in our evolution. It's time to return to thinking of intelligence as a collaborative act, not an individual's test score. The range and complexity of problems that we face, from the climate emergency to global water and food shortages and the threat of the next pandemic, mean that we need all brains on deck. We must develop ways of collaborating across groups of people with different perspectives and experiences from our own. We need to value a collectivist approach to intelligence and understand how it emerges, what skills it depends on and which activities in the brain drive those skills.

We can all think so much more intelligently about our interactions with one another, and our approach to our biggest challenges. When we make this shift, from 'me' to 'we' thinking, our worldview changes, our imagination is unleashed and every single one of us is able to contribute our unique viewpoint to humanity's pool of intelligence. It's exactly this exhilarating joined-up thinking that we need now. Let's harness our collective brain power and see how far it can take us.

CHAPTER ONE

The Power of Joined-up Thinking

Neuroscience has been investigating intelligence for decades but until very recently it has treated the brain as a single entity. A lot of work has focused on understanding how nature (our particular brain, built on the blueprint of our DNA) interacts with nurture (the experiences we learn from, especially in early childhood) to give rise to our consciousness and our unique experiences.

It's only in the last few years that neuroscience has broadened its focus. It has shifted away from studying brain regions as separate areas with specific functions, to treating them as a network of staggering sophistication: the connectome. Now, scientists are looking at how intelligence arises within the brain–body system as a whole, and between a group of minds that influence one another. Did you know, for example, that electrical oscillations between people's brains synchronise when they're engaged in a communal activity, so that people are more likely to literally see the world in the same way? This boosts our capacity to learn together or to build a consensus. But during periods of stress, fear and conflict the process can go awry, undermining the extent to which ideas can hop from brain to brain and become seeded in a collective way of thinking.

The ongoing revolution in brain-imaging technology allows scientists to study the brains of living, breathing, learning, interacting creatures in detail. This means that neuroscience is increasingly able to illuminate the way we think and behave. Cognitive scientists can now investigate even highly abstract behaviours such as compassion and guilt. They are also delving into the way brains collaborate, looking at how a group works harmoniously to solve a problem and what happens in people's brains to enable this collective success.

A lot of the cutting-edge research now coming out will revolutionise how we think about intelligence. Studies into embodied cognition are investigating how we can tap into the vast amounts of information stored in our bodies, much of it picked up unconsciously from signals given off by other people. The interface between artificial intelligence and human intelligence is another space of pioneering exploration for cognitive scientists, and science-fiction-type innovations are coming thick and fast. Memories can now be electrically imprinted from a donor's brain into a recipient's, opening up a route towards the possibility of downloading expertise. Brainets are being created, in which brains are wired up to allow individuals to collaborate through direct brain-to-brain transfer of information.

But to my mind, some of the most exciting conclusions are less futuristic and much closer to home. The studies that show how healing from anxiety and distress can be a collective endeavour, for example, or how essential it is to break down dominance dynamics so that good ideas can emerge and seed themselves in a group.

Collective intelligence is the stuff of our everyday lives and we all do it, often without even noticing. Every time we diffuse a family row, organise a big social gathering or collaborate on a project at work, it comes into play. Family life, social life, working life – all of them are built on collective intelligence. The cognitive skills it depends on are as much emotional as they are analytical. They involve communication, trust, empathy, persuasion, negotiation, imagination, wit, emotions and language. Collective intelligence flows from one brain to another, morphing and enriching itself as it goes to create an extended mind that transcends any individual brain. A mind that's infinitely smarter than any single one of us.

Collective intelligence emerges and flourishes in certain conditions, as we will see. A fundamental precondition is of course social connection. Without contact between people, preferably real-world contact, there can be no joined-up thinking. Nothing could make the point more clearly than the lockdowns imposed in response to the coronavirus pandemic. Social isolation increased and then fluctuated over a period of months, as countries came in and out of lockdown. This created the ideal conditions for neuroscientists and psychologists to study numerous aspects of human behaviour, including those that influence intelligence.

Not that you needed to be a cognitive scientist to observe the impact of lockdown on our ability to focus, think or communicate. Many of us experienced a sort of brain fog that combined feelings of low mood, mental exhaustion and distraction. Studies conducted in Scotland and Italy, among others, measured cognitive functions as the pandemic progressed, observing that lockdown periods coincided with a fall in people's brain power, which picked up as restrictions on social interactions eased. Those who shielded improved more slowly.

There is a heap of data from long before the pandemic to support the assertion that we all benefit from interaction with others; that in fact, our most fundamental skills depend on it. Human beings are social animals. Our physical and mental well-being, cognitive functioning, language acquisition and emotional regulation all develop in and depend on an open and diverse collective life. Our thinking power becomes greater when we are part of a group that includes individuals from outside the family. We all need exposure to diverse role models and the perspectives of people beyond our genetic kin and our immediate living and working situation. When we are in communication with ideas and people from beyond our bubble, our individual cognitive abilities increase and we can contribute our unique perspective to the collective mind.

Luckily for us, we've created a dazzling array of technological tools to support the connectivity we need in order to function, even during periods of physical isolation. Increasingly we live in a world where the collective mind no longer depends on interactions in real life but exists on internet-enabled platforms. We can share ideas and opinions across the globe, in nanoseconds. We have access to all the information in the world.

The interplay between technological development and social change is a constant of our species' intellectual progress. We come up with new tools to investigate and implement the new ideas we're having, and those tools then create new possibilities, which drive further cultural and social evolution. It fascinates me that over the last thirty years, we have been driven, both scientifically and socially, to develop technologies that allow us to observe and utilise our connections – both between brain cells and between brains. Magnetic Resonance Imaging (MRI scans) and high-powered electron microscopy have shown us the nerve tracts and synaptic connections that link up regions in our brains. New media and internet-based technology facilitate information exchange between us. It's as if we've created the perfect environment to jump-start our evolution towards the joined-up thinking we need.

This seems to be precisely what's happening. Studies undertaken by cognitive scientists between 2020 and 2022 suggested that for the first time in human evolution, our environment – this techno-enabled, networked, communication-driven landscape that we inhabit – is now directing our species' evolution at such a rate that its effect supersedes even the value of our genes. Our group-level cultural evolution is now more adaptive and more rapid than genetic evolution, which means that our environment – the various cultures we grow up in and inhabit – is becoming the ultimate force that shapes how we as a species will progress.

This concept is the antithesis of what I was taught as an undergraduate studying cellular and molecular biology, and, if confirmed, it will constitute a completely unprecedented shift in humanity's cognitive development. As Timothy Waring and Zachary Wood concluded in their 2021 Royal Society review, 'If genes hold culture on a leash, culture is dragging them straight off the trail.'

Why is this so important? Well, it's thought that big evolutionary transitions occur when groups have developed the ability to cooperate so well that competitive selection between individuals becomes less important. One such transition drove the evolution of life as we know it today. When the individual cells that arose on Earth around 3.8 billion years ago developed enough ability to communicate, cooperate and integrate, eventually, through trial and error, they assembled into complex biological organisms – including us.

Our species may be entering another evolutionary transition, where our collective intelligence can begin to properly evolve and emerge from behind the individuality of our past. Might we be about to enter a new era of development, evolving towards becoming a socially integrated mega-group, much like beehives and ant colonies? This concept might feel alien, but it could usher in a utopian era of human cooperation.

Meanwhile, there's no shortage of excitement and positivity in the era of the brain in which we already live. The mesmerising cartography of our minds is being revealed in ever more detail through MRI scans. This allows us to observe the natural breadth in thinking styles offered up by the range of human brains. We are starting to appreciate the strength of neurodiversity among our species, and how it can benefit us all. A person diagnosed with autism has a brain that perceives the world in a different way from that of a person without, or one with a diagnosis of dyslexia. The brain of a teenager is structurally different from that of a pensioner. These distinctions have profound implications for cognitive style. If we can value and capture the diversity of thinking available to humanity, how much greater could our problem-solving and creative capacities be?

The dedication page of this book features a piece of art entitled *This Place*. It was painted by Alicia Adams, a proud Kamilaroi woman. The Kamilaroi nation is of vast expanse and the second largest nation on the east coast of Australia. The dots represent the saltwater and freshwater people, different clans coming together to collaborate and celebrate their creativity. It depicts collective intelligence forming, from a bird's-eye view. At the centre the tribes merge in a place for storytelling where they share their perspectives. Out of this, new knowledge arises. Space is also made for reflection on historical wisdom so that it can be passed down the generations and incorporated afresh into their ongoing thinking.

I love this image. Alicia's passion for the wisdom held within her community resonates so strongly with my own scientific knowledge. Her dot paintings recall the maps that researchers construct to visualise the constant flow of data across the brain's connectome, which functions as the storytelling machine that generates the unique narrative of our lives. For me, Alicia's image is like one of these maps but on a larger scale. Rather than each dot representing a separate brain region, it stands for an individual person. The picture describes the interactions between people that add up to the story of a community. They are a portrait of collective intelligence at work.

Let's start our exploration of this new approach to intelligence by looking at how our brains develop to work together within our family units. The family is the cradle of collective intelligence. It is the first group we join and the smallest group in which most of us live. From our birth families we inherit our genetically determined capacities and dispositions. We also learn early lessons about how to deploy them: how to think and how to behave. The family is the perfect context to ask questions about how an individual's intelligence emerges, interacts with and is influenced by that of others.

From there we will move on to look at bigger, more diverse groups of people, such as those who come together at work. In these groups we interact with people who are not genetically related to us, so there is a greater diversity of cognitive skills and points of view. This has the potential to increase collective intelligence but also presents challenges because as the group gets larger, there is a risk that perspectives can go unheard and conflicts can arise. What are the skills and behaviours we can use to get round the challenges and be successful? How can we embed intelligent strategies for leadership and collaboration?

Throughout the book we'll be expanding the size of the groups we're thinking about and looking at collective intelligence in ever broader terms. How do different tribes treat each other in ways that either foster collaboration, or trigger a collapse into conflict? Are there ways to scale up the positive social skills that underpin collective intelligence or do they inevitably get lost in the competing crowds?

Social life has to a great extent shifted online, and we will look at how it plays out (both intelligently and not so intelligently) over the web. Human beings are driven to share ideas and learn from each other but we are also vulnerable to manipulation and disinformation. Sometimes groups encourage each other's limitations and prejudices, stifling debate and ramping up attacks. This can have potentially serious consequences, not just for problem-solving and the exchange of ideas but for people's safety and wellbeing. We should not underestimate our species' capacity to behave in profoundly self-defeating ways; but evidence shows that the best way to steer away from them is to practise embedding the positive social and emotional skills that underpin all collective intelligence.

In order to tackle the really big tasks that challenge us, we need to be able to harness huge amounts of cognitive capacity. Where are the inspiring examples of ambitious joined-up thinking that spans sectors, countries and even generations? Can we learn from our ancestors how to build flexibility and resilience into our thinking and how to have faith that the generations who come after us will complete the projects we start? If we can learn how to be good ancestors ourselves, we will maximise the chance that we pass on a positive legacy of social and emotional skills that will enable our descendants to flourish.

Perhaps artificial intelligence could give us the brain boost we need to take on these enormous tasks? The intersection between our human intelligence and artificial intelligence is getting more intimate all the time. Communications technology and neuro-technology are evolving in tandem and in dialogue with one another. Can we embrace this new strand of diversity or should we fear it? And what can it tell us about our limitations and possibilities as intelligent, empathetic, creative beings?

The starting point for this journey is a dive into what we know about intelligence. What is it? Where did it come from? How do human beings' intellectual capacities compare with those of other intelligent creatures? If we begin to think of it not so much as an individual's intellectual capacity but as a shared survival strategy that makes us fitter for life, we will be ready to reimagine it for the challenges of the twenty-first century, enabling us to thrive in a hyperconnected world of increasing uncertainty and escalating change.

I believe we are now at a tipping point where we can perceive the limits of individual intelligence. Now is the time for a renaissance in joined-up thinking that harnesses the diverse cognitive reserves at humanity's disposal. If we can nurture the combined brain power of the many, across groups and across generations, by opening up to ancient wisdom, to intellectual mavericks and outlier ideas, we can shift from 'me' thinking to 'we' thinking. This is the mindset that will drive our success, both as individuals and as part of our many collectives, over the next crucial decades.

CHAPTER TWO

What Is This Thing Called Intelligence?

Most of us associate intelligence with certain skills (mathematical reasoning, say, or being able to speak a number of foreign languages) and also with concrete achievements in the form of test results, discoveries, innovations and prizes. As children, our history lessons focus on the stories of outstanding individuals, the Marie Curies and Charles Darwins, the Mary Annings and George Eliots. We know of course that geniuses are by definition exceptional, but we usually accept that some of us are just provably, measurably smarter than others. We learn this at school as we experience streaming for ability, and sit exams. By the time we arrive at adulthood we have absorbed a whole set of beliefs about what intelligence is, what it produces and what it looks like. We have grown up in a society that is heavily invested in a hierarchy of cleverness, and the need for institutions such as universities and corporations to define, develop, test, monetise and reward it.

In this model of thinking, where our individual intelligence is equated with success at school, college and work, intelligence stems from the innate cognitive gifts we were born with and is then developed through education, measured with exams and finally presented with opportunities to prove itself through innovative products or ideas. Intelligence becomes a competition, with winners and losers.

Now, I admire conventionally intelligent people as much as you would expect from someone who is lucky enough to work alongside some very brainy individuals. There's no denying that some people have unusual gifts, and I am grateful for all the contributions they have made throughout human history. But as a biologist and a neuroscientist, I also know the perils of over-investing in any single individual trait – or individual person. When we conceive of intelligence in narrow terms, we are falling into that trap.

Our species has thrived because of diversity (as have all other social animals). Any and all contributions might prove to be crucial to solving a problem. Investing too much in the usual suspects – the high-achiever or the lone genius – can blind us to the range of skills and capacities that other people have to offer. The breakthrough might come from the introvert who has a completely novel approach to a long-standing challenge but lacks the communications skills to present it. Or from the employee who's considered too young to be up to the job, or the one who's reckoned to be past it, or the person with a diagnosis of attention-deficit/hyperactivity disorder (ADHD), whose creativity and lateral thinking could generate ideas for new products and approaches.

I'm not for a second saying that conventionally intelligent people or experts aren't needed any more, just that broadening our definitions of talent and expertise will yield more break-throughs in every area of life, from science to our personal relationships. We limit ourselves and others when we fall back on our ingrained thinking about how to be skilful, successful and smart. We need to find new answers to the question of what it means to be intelligent, by pushing beyond our immediate associations with certain skills and certain kinds of people. That will free us up to imagine intelligence anew and, from there, figure out how we can build more of its variety into our own thinking and into our groups' interactions.

All Brains Are Not Alike

Human beings have always been proud and possessive of our intellectual prowess. For centuries Western thought rested on the belief that we alone of all animals were conscious, thinking creatures. But just as people have always admired and desired intelligence, so too have they questioned what it means to be clever. Discussing, let alone measuring, intelligence has always thrown up questions. Is it an output or a process? Is it innate or can it be taught? Is it flexibility of mind, capacity to reason, flair for creativity or something else entirely?

Most people would probably agree that a high IQ or a clutch of good exam results can tell you *something* about a person's abilities but doesn't capture what we understand by intelligence in a broader sense. A straight-A student may have a superb memory and excellent powers of analysis, but how's their emotional intelligence? Do they come up with unusual and original insights? Are they witty? Are they shrewd, adaptable, a quick learner, curious? Do they have good social skills? Do they empathise, and communicate well? What exactly are we talking about here? And beyond IQ tests or exams, can we judge intelligence by the quality of someone's conversation or by their life choices?

The answer is that we can and do make such judgements all the time, but they are subject to our own partial and biased notions of what it means to be an intelligent person. If we rely on a set of A grades as proof of intelligence, how do we accommodate the fact that coaching can significantly increase a student's chances of achieving those grades? If we define it, consciously or unconsciously, as a set of accomplishments and tastes, can we recognise that such judgements are subject to our bias about social groups? At various times over the last hundred years, IQ tests have been used to justify the belief that certain races are inherently more intelligent than others. This has been soundly disproven over and over again but it shows us that a little bit of science, selectively applied, can be used in the service of almost any argument. Intelligence, even in the narrow sense of an IQ score, is less a measurable fact and more a label that we use to define and value certain qualities.

My discipline of neuroscience is naturally the main framework for my own thinking about thinking, and neuroscientists are always looking out for the divergences between what brains *generally* do and what particular brains do differently from one another. Neuroscience has its sights trained on what a baby's brain can do and what it can't, yet; or what a person with schizophrenia's brain does, compared with the brain of someone who doesn't have that diagnosis. This focus on difference and diversity feeds into the definition of collective intelligence that we will be exploring throughout the book.

In functions that underpin a traditional view of intelligence – such as short-term memory and problem-solving – neuro-scientists have demonstrated that certain groups have physiologically different brains, with different functionality. Information processing slows down with age, for example, and the older brain is more vulnerable to bias because it relies on stored wisdom (or ingrained ideas!) to compensate for slower speed. The teenage brain has fewer connections between regions, making it harder for teenagers to integrate reasoning and emotion. Consequently they are prone to impulsive decision-making, but also to coming up with new solutions to problems. The brain of a person with ADHD is more sensitive than average to the rewards of novelty, which can fuel high levels of curiosity.

These differences are subject to a range of variation according to the individual's particular neural circuitry, but they are observ-able differences in the physiology of the brain that pertain to particular groups. These are what we call *cognitive* diversities.

Then there are *social* diversities that arise from a person's ethnicity, nationality or social class – from nurture not nature, as it were.

The brains of people with different backgrounds are not structurally different from one another at birth, but diversity of experiences can give rise to differences in ways of seeing the world and processing information. This is because the brain is constantly updating its neural networks in response to new information. Plasticity (the brain's capacity to change at the level of its synapses – the connections between individual neurons) allows us to learn new skills and process new information. It gives rise to our learned behaviours, from simple task fulfilment such as knowing that we must turn left then right to reach the office, all the way up to highly complex behaviours such as our use of language in different settings, or our response to our emotions.

Learned behaviour is a crucial element of how we need to think about intelligence in a broader way. It can give rise to significant differences in cognitive styles, which occur when thinking has become habitual. This can happen to people with specialist skills and knowledge, such as cab drivers and academics, who repeat tasks so many times that they become experts. Particular styles of thinking also arise in different social groups such as men and women, Black, white and Asian people. A lived experience, if it happens over and over again to an individual, can give rise to a particular way of seeing the world and processing information.

Throughout the book we will be looking at how these and many other varieties of cognitive and social diversity can be valuable for the individual, because they represent an adaptation to their particular circumstances, and can also bring valuable diversity of perspective, cognitive style and brain power to the collective.

Beyond Reason – For a Diversity of Intelligences

As well as showing us how aspects of conventional intelligence arise, neuroscience is now exploring (alongside behavioural and social psychology) how other crucial skills and capacities develop.

We will be looking at emotional intelligence, which is underpinned by empathy and compassion. We'll be examining the science of intuition and exploring how gut feelings arise. We'll be looking at the crucial importance of communication, listening and turn-taking for effective collaboration, and examining the pro-social behaviours and modes of thinking on which these skills depend. Broadening our thinking about thinking in this way can open our minds to different skills that we could make more of. Few of us know how to cultivate the full range of our own capacities, let alone the spectrum of thinking available to us as a group. There is a whole world that opens up when we begin to explore intelligence more broadly.

Throughout the book I will be defining intelligence not just as the ability to recall or interpret information or to determine the next figure in a sequence but as the capacity to solve any problem efficiently and effectively, at speed and under pressure. When we think in this way we immediately see that there are almost as many ways to describe people's problem-solving abilities as there are problems and people. We begin to see evidence of intelligence in somebody's capacity for negotiating; for reinventing themselves in a new role; for healing after trauma; for resolving conflict; building a community; leading an organisation. There are many problems to solve in this complex world of ours, and many ways to approach them – even as an individual, let alone as a collective.

One way of broadening our thinking about intelligence is to consider it as an evolved strategy that tends towards the success of the species, as well as the individual. There are certain aspects of intelligence that almost all of us share – our species' incredible inheritance of reason and language, creativity and agility – and then there are particular adaptations. It's the power and value of this diversity that I want to look at in more detail.

Why Groups Are Smarter

Intelligence is a survival mechanism of staggering sophistication. In human beings it has evolved into a system of intersecting behaviours that, when they are deployed skilfully, make a successful outcome more likely. Whatever the field or the goal, whether it's developing a new product or crossing a road safely, the skills that make success more likely will have similar biological underpinnings. They will be in the interests of our species' collective long-term survival, written into our brains through evolution, and coded for by our particular DNA. There's no denying that us human beings have evolved to be, generally speaking, incredibly clever.

That said, many of us still fail to understand our limitations. We struggle to grasp that the human brain is not a single entity that serves our unified being but a staggeringly complex electrochemical network that, though infinitely sophisticated, is also fundamentally flawed. Perception, for example, relies on your brain working round the clock to interpret your surroundings and put together a model of reality for you to inhabit.

This is a vast job. It's been calculated that a whopping 11 million bytes of data are sent to your brain *every single second*. Signals are picked up by your sense organs and turned to electricity as your brain pumps sodium and potassium ions into and out of its 86 billion or so nerve cells. The resultant dance of data zips across your brain network at speeds of up to 250mph via the 86 trillion or so synapses that connect neurons to their neighbours, to form the most intricate and complicated circuit board imaginable. Each cell plays its small part in the processing that generates your eventual behaviour.

It's a phenomenal feat but it's a rushed job, and the brain makes errors as it goes. Perception is not a matter of the sense organs recording reality and the brain interpreting it, but something much messier. Neuroscience has shown that all of us, however clever we may be, are vulnerable to the same cognitive errors in perception and decision-making. We're limited and we're biased, because our brains have so many jobs to do that they must rely on short-cuts, deferring to interpretations that have served us in the past rather than evaluating whether the situation is different this time. We prioritise some information and ignore other bits. Some signals get dumped (unconsciously) into the bin marked 'unimportant trivia'. Errors can creep insidiously into our decision-making and opinions. We jump to conclusions, defer to authority, conform to our peer groups and even miss the gorilla on the basketball court if our attention is directed elsewhere. (Literally, as we'll see in the next chapter.)

The limitations of our individual brains provide one of the most compelling reasons for why we've evolved with the capabilities and inclinations to support collective intelligence: the so-called 'pro-social behaviours' (empathy, altruism, effective communication). They are the key skills that underpin collective intelligence. Two brains really are better than one for arriving at a reliable and objective understanding of a situation, precisely because two people can, to a certain extent, correct each other's perception errors and biases, and negotiate their way towards the most robust interpretation of whatever they're engaged with.

Most of us are pretty invested in thinking of our particular brain as a smooth operator, even once we know the extent to which our thinking is full of holes and riddled with bias. We imagine our brains as the seats of our consciousness and the factories of our identities, and though this belief may be a fiction of our egos, it's certainly a useful fiction. Life in all its rich emotional, sexual, cultural significance would be less interesting without the idea that I am me and you are you and we are not just different, not even unique but special. (We human beings have big egos to go with our big brains!)

But while it's both necessary and valuable to be invested in our ideas of selfhood, there is much more to know about ourselves and our collective identities and capabilities. If we can see past the 'I' to the 'We' that lies beyond, our understanding of ourselves gets more nuanced, not less. Collective intelligence doesn't take anything away from our individual intelligence. It's not an either/ or situation so much as a widening of perspective. We are extraordinary and unique individuals *and* we are collective organisms, both at the micro level of our bodies' organs and the macro level of our need to live in contact with other people.

Survival Is All about Networks

In the early part of the twenty-first century a new field of biology emerged that focused on the way cells interact as part of a complex system. This system was termed the 'sociome', which rather anthropomorphises molecules and cells, but it's an evocative description of their ability to communicate in such complex ways that they might almost be said to be 'socialising'. Since then it's been increasingly accepted by biologists that all life, at every level, is social: from genes cooperating to form organisms, to animals cooperating to form groups.

Every element of our bodies depends on connected networks functioning across groups of cells and even across distinct organisms. Our guts host a microbiome of millions of microbes, which have a huge influence on health, happiness and – you guessed it – intelligence. There is a society of biological processes occurring in each and every organ of our bodies, which is in turn connected with multiple others. Our brains are connected to our peripheral nervous systems and in constant communication with the heart and the gut, which are both rich in nerve cells and keep hold of a lot of the extraneous data that the brain can't process, storing it for possible use in the future. A thick cable of nerves called the vagus nerve connects the gut to the heart to the brain, rooting through to the insula, a brain region involved in forming our perception of the world. All of which helps explain that 'gut instinct' or 'feeling of the heart' when we suddenly feel we know something, almost unconsciously.

Professors Sarah Garfinkel and Joel Pearson, at University College London and the University of New South Wales respectively, have discovered that for some people this connection between the organs is stronger and more accessible to their awareness than it is for most of us. These individuals are highly sensitive to information they're receiving subliminally. Their intelligence is optimised by the power of what neuroscientists call 'embodied cognition' so that it's almost as if they're thinking with their guts and hearts as well as their brains. They have access to a kind of collective intelligence that emerges from within their own bodies. We'll come back to this later and look at the evidence that we can all develop our own internal collective intelligence by tapping into our embodied cognition to develop our intuition.

As well as the connections between cells and organs that determine our health, happiness and intelligence, evolution has also prioritised connections between individual people. Essentially, human beings have evolved to live in groups. Faced with stronger, faster predators or the constant work of childbearing and food gathering, an individual or pair bond of human beings was vulnerable. Our ancestors relied on one another in multigenerational families and tribes, for support to bring up their children and provide the basic necessities for survival. They developed communication in order to share ideas, and empathy so that they could, when required, prioritise the needs of another person or the group over their own. Without these propensities to communicate and look out for one another, our species may not have survived and almost certainly would not have thrived as we have. The individual benefited too, because a person whose skills were of value to the group was powerful and secure.

You could say the same things about any social species, mind you, and when we start to drill down into sociability it turns out that it's everywhere, from ants to honey bees, and even forests. Collective intelligence in all manner of organisms arises from social interactions, and it serves the needs of the group while reinforcing the status of the individual. We're not so different from ants in this respect.

Understood in this way, collective intelligence is part of our evolutionary inheritance and a superpower hiding in plain sight. Rather than over-relying on our individual flawed brains or even the individual flawed brain of a high-achiever, it's time to learn from the way things are in the rest of nature.

Intelligence Is Always Collective

Here's a challenge to our species-centric beliefs about Homo sapiens' abilities and virtues: even brainless plants exhibit a degree of collective consciousness and sense of community. Trees emit chemical signals to warn their neighbours of potential threats, communicating that they should mount their defences when predators are lurking or infectious agents are passing through.

David Haskell, professor of biology at the University of the South, has been nominated for the Pulitzer Prize for his work exploring the community that emerged in a square metre of forest over a year. Speaking to *The Atlantic* magazine about the cognitive capacity and community spirit in our woodlands, he said, 'I'm very comfortable using words like intelligence, but I need to emphasize that this is a very other kind of intelligence . . . We're not imagining one big super-organism that thinks in a human-like way. To me, the closer analogy is with human culture . . . It's very decentralized, but it has memory and contributes to our understanding and our ability to solve problems.'

Most organisms on Earth have evolved as social beings, and

for good reason. Animals create tight-knit communities, whether they be shoals of fish, flocks of birds, swarms of bees or colonies of ants. Ant societies exhibit great collective intelligence, running communal farms where they tend aphids for 'milking', and take turns to defecate on strawberry pips they've secreted in crevices to nourish the seedlings. Birds flock together partly to boost their collective intelligence. A murmuration of starlings flying in a winter sky at dusk is a mesmerisingly beautiful sight. The roost can swell to around 100,000 members, diving, swooping, flying in precise synchronicity. Computational mathematicians have speculated that each starling subconsciously replicates the flight path of its immediate seven companions and it's this mimicry that gives rise to such impressive harmony. The birds also work collectively, sharing information about food sources. When food becomes scarce, roost numbers actually increase. The benefit to greater cognitive capacity, which can be used to discover new sources, outweighs the cost of increased competition for them.

Bees may have a clearly designated leader but communication among all hive members informs the important question of moving home. When a hive becomes overcrowded, it splits in two. The moving colony sends out a few hundred designated scout bees to explore the terrain and choose a location for their new home. When the scouts return they deploy a complex series of figure-of-eight dance moves to communicate how much they rate their spot and its exact location. A surprisingly democratic decision-making process then follows. More scouts go out to verify the possibilities, and there might be a bit of bee headbutting if it becomes a tight contest for sites. But eventually a consensus is reached and the bees follow their new queen to the selected site.

When we look to nature we see time and time again that individuals work collectively and the group is much smarter than the sum of its individual parts. Communities create a chamber where intelligence can amplify, where the decisions made together are generally much more effective than those made alone. Communication and cooperation are evolved strategies adopted by all living creatures. They boost survival chances, bring social benefits and compound effectiveness at problem-solving.

Collective intelligence of this kind is not sentimental, mind you, in either humans or ants, and sociability is not without downsides. Living in high densities increases the risk of disease transmission and competition for resources. But even so, its benefits outweigh its costs. Cooperative or pro-social behaviour, understood in biological terms, is less about the milk of (human? ant-ish?) kindness and more about maximising our individual and our species' survival.

Members of our species may occasionally headbutt like disagreeing bees, but generally we are well equipped to work collaboratively. Robin Dunbar, professor of evolutionary psychology at Oxford University, has spent his career researching the social brain and says, 'Most people, even the relatively shy or introverted, have a staggering capacity to navigate our complex social world, and much of this functioning takes place with seemingly unconscious ease. This suggests that all relationships are to some extent reliant on deep-brain functioning to do with pleasure, reward and motivation.' Human beings seem to be biologically driven to find the process of reciprocating attention rewarding, and Dunbar suggests that this pleasure in communication and collaboration has helped us to evolve as a species.

Bottom line: whether we are an ant, a tree or a human being, as individuals we are vulnerable. In our species, the social neural circuit, which is not so much a brain region as a labyrinthine system, has evolved to offset this vulnerability. It drives us to form connections with other people that will come in handy in times of crisis and buffer us against loneliness, sadness and ill-health.

Humanity's Unique Contribution to Collective Intelligence

Effective collaboration on any kind of project, whether it's farming aphids or building a space station, is the output of an evolutionary strategy that has prioritised the necessary skills for working together as a group. But Robin Dunbar's research has shown that the size of the group matters. In humans, once it passes approximately 150 individuals, cohesiveness starts to break down. The bonds of trust and reciprocity on which collaboration depend cannot be maintained. Group size varies from species to species but all species have an upper limit beyond which they must break away and form a second group. This places a limit on the amount of brain power that any group can harness.

Human beings are the only animal, as far as we know, that has come up with mechanisms for getting round this limitation. By developing rules to govern interactions, as well as institutions to oversee them and arbitrate over differences, we have been able to massively extend our networks and so work effectively in ever larger groups. This has been invaluable to our species' development. If we were only able to rely on straight reciprocity – you do something for me and I'll do something for you – then our cooperative world would be far smaller, probably confined to our core circle of family and significant friends. There would have been limited possibilities for trade, commerce, travel or culture to develop. Human activity would have remained tribal or feudal.

Nichola Raihani traced the development of social rules and institutions in her book *The Social Instinct*. As far back as the Pleistocene era there was a shift away from the winner-takes-all dynamics of a strictly hierarchical group structure, and towards governing by coalitions. This helped our species to flourish. In the kinds of hierarchies we see in other primate groups such as chimps, a single high-status individual battles it out with others at the very top. Chimps do form strategic alliances around the alpha but since there is only one winner the rewards are highly concentrated. By joining a team or becoming part of a syndicate, the chance of receiving some sort of reward is much better for many more individuals. There's an incentive to collaborate, especially in a challenging environment. As this approach beds in and it makes sense to more and more individuals to work together, so the skills and propensities that drive such collaborative behaviour are favoured by selection.

According to Raihani this process has continued throughout human history, from the guilds of the Middle Ages to our own digital era where we have developed tools such as Trustpilot or Airbnb's ratings, which increase transparency in online interactions. All this codifying of group behaviour and finessing of the skills needed for collaboration drives the emergence, and evolution, of collective intelligence.

The prospect of humanity evolving into a super-organism with a collective consciousness may be a long way off but it's not an impossible scenario. In the meantime, we are definitely experiencing a surge of creative collaboration in every area of human endeavour, from science to politics and beyond. Digital communication, with its infinite interconnectedness, has driven an explosion in our capacity to share our ideas and work together.

Eminent biologist Richard Dawkins coined the term 'meme' back in 1976 in his groundbreaking book *The Selfish Gene*. He described memes as 'units of cultural transmission' – ideas or images or pieces of behaviour that were particularly ingenious and apt for a situation. Memes, like genes, seek to spread as far as possible: in the case of memes, either through language or by direct observation. The new meme competes with existing ones and successful memes displace less robust ones, mutating as they circulate in different contexts. In this way, our thinking about the world evolves over time and in different locations.

With each revolution in communications technology from

the invention of writing onwards, the speed and range of a meme's travel has been increasing. Malcolm Gladwell's *The Tipping Point*, published in the year 2000, described the way cultural trends typically grow slowly until they hit a certain number of participants and then spread like wildfire, implanting themselves in mainstream culture. The book appeared before Google, social media and smartphones had transformed our world and our lives. Since its publication, of course, the speed at which messages can reach billions of people has exponentially increased. Memes can be communicated in milliseconds, hopping oceans and languages, generations and cultures. In this digital environment, knowledge is becoming less powerful for being easily searched and it's memes that are rising in value, since they represent a creative response to that knowledge and an attempt to solve a problem, whether big or small.

The internet has clearly not turned out to be the utopia that its early adopters hoped for, but there's no doubt that it is driving a renaissance in joined-up thinking. Although not all its applications are positive for society (and some are definitely negative), there are many examples of diverse and novel thinking being pooled to tackle problems at scale. From citizen-science projects that recruit snorkelling tourists to upload their underwater holiday snaps, via Extinction Rebellion and on to crowdfunded research into psychoactive drugs, there has been an explosion of collective intelligence operating online. The problem might be the threat to coral reefs, failure to tackle the climate emergency or the need for new approaches to treatment-resistant depression. In every case, the internet has enabled groups of diverse people to gather, discuss, raise money, come up with strategies and organise themselves to take action as a collective.

Whatever we think about the merits of these projects and causes, we can see the same skills and structures being deployed in order to create fluid, non-hierarchical, open, trusting, collaborative interactions between people with diverse points of view.

These systems are not perfect and are not bound to succeed but they offer us the opportunity to learn and experiment as part of a bigger entity than ourselves.

The Age of the Extended Mind

Our capacity to reason, invent, discuss ideas and come up with new solutions to our problems is now fundamentally shaped by being able to access other people's thoughts and ideas in detail, at scale and at speed. The internet has joined up all our thinking whether we personally engage with it or not, because it has altered the values of information exchange to such a massive extent that the landscape of our thinking has been transformed. For philosopher and cognitive scientist Andy Clark from the University of Edinburgh, this has philosophical implications as well as offering us a tool of infinite possibility.

In 1995, Clark published a paper jointly authored with David Chalmers, his long-term collaborator, entitled 'The Extended Mind'. It explored the way our particular minds are not confined to the physical matter of our brains but extend beyond our bodies to interact with other people's minds and also with the tools we think with, such as the pencil we use to scribble notes. According to this line of thinking, the mind is not something that sits inside our skull, reducible to the electro-chemical processes of the brain or even to the complex inter-actions between new information and our ingrained cognitive habits. The mind is always reaching out into the world and connecting with whatever it finds there. It is a process. It's what happens when you and your colleague spark ideas off each other or you and Google Maps figure out the best route for the road trip. Even language, Clark argues, is the incorpo-ration of a tool into the mind. Out of this rich ecosystem, intelligence emerges.

Since the publication of that paper, smartphones have replaced notebook and pen, and ideas that once seemed outlandish have come to seem obvious. Clark and Chalmers were ahead of their time in anticipating that thinking could occur across brain-to-tool and brain-to-machine connections. Their insights were at least as metaphorical as factual, posing the age-old question of what is a mind in the context of cyborgs – brain–machine hybrids.

Nearly thirty years later, the concept of the extended mind has come to feel like a description of our infinitely joined-up world and our position in relation to it. The cyborgs that so fascinated Andy Clark in the 1990s aspired to hack their brains as well as their arms in order to take on the capacities of machines. We can upgrade our own intelligence by less drastic measures, through incorporating the salient knowledge that we need in order to make our brains fit for twenty-first-century purpose.

Our first site of investigation into the practical detail of how joined-up thinking happens is the nuclear family. Our birth family is the first group of people any of us joins. For the majority of us brought up in that context, our parents not only give us the blueprint for our brains but also teach us our first life lessons. The family is truly the cradle of all learned behaviour, and all forms of intelligence.

CHAPTER THREE

Family: The Cradle of Joined-up Thinking

Those of you who are now or ever have been the parents of small children will at some point probably have enjoyed (or perhaps endured?) a family Lego session. So you will recognise the situation in which I found myself with my four-year-old son, Max, one rainy afternoon during the lockdown of March 2020.

We had got stuck in Australia as borders closed and coronavirus rampaged across the world. We ended up in Queensland, having completed the last leg of my book tour, and found ourselves on a sort of indefinite holiday, knowing nobody. We were trapped in paradise and there were many things to love, despite the initial isolation and uncertainty. But our extended family and friends were back in the UK, flights were grounded indefinitely and we hadn't yet met the amazing people who would go on to become almost like adopted grandparents and siblings to Max. On this particular day of tropical storms. we couldn't play outside and were housebound. So naturally, our thoughts turned to Lego.

Max and I fall firmly into the category of chaotic Lego builders. Not for us the tidy-minded strategies of sorting pieces into colours or keeping kits separate. That afternoon, we tipped over an overflowing box and I stared, feeling somewhat overwhelmed, at the thousands of pieces scattered across the table, while Max enthusiastically waved the instruction booklet for the fire engine.

We quickly settled into a rhythm: I searched among the blocks for the correct pieces and handed them over for him to configure, step by step. After a slow but happy half an hour, I came up against a problem. I couldn't find the particular piece required for a corner of the engine's roof – an innocuous grey 2x2 plate. I'd been searching for ages, increasingly irritated. 'What's up, Mama?' asked Max, in response to my huffing and puffing. So I told him that we were stuck. His response filled me with parental pride, tinged with embarrassment at my own intellectual shortcomings.

Max solved the problem immediately, handing me a red tile that had been sitting right in front of me. OK, it wasn't grey, but the shape and size were perfect. I'd been using colour as my initial search term and had failed to spot the obvious: that it didn't actually matter. The more crucial factors were size and functionality. After that breakthrough we romped through the rest of the kit. Later that evening we proudly showed off the fire engine over WhatsApp to Max's grandparents back in the UK, complete with a couple of tiny insignificant patches of the 'wrong' colour in an otherwise perfect construction.

Our family Lego session provides a snapshot of collective intelligence at work. My older brain is physiologically very different to Max's rapidly developing brain, and as a consequence we have different problem-solving capacities. It's typical that Max was able to come up with a workable solution to the problem of the missing piece when I could not. He could think laterally by undertaking a so-called 'conjunction search', outside the initial search parameters. In other tasks his younger brain would be 'wasting' energy by carrying out this kind of work and my older brain, better able to stick to the brief, would be more useful.

My inability to spot the solution in plain sight is an example of a phenomenon called inattentional blindness. The authors of one of the original studies into this, Daniel Simons and Christopher Chabris, were based at Harvard University when they asked people to watch a video of a basketball game and count the number of passes around the court. Most people did pretty well at that task

– but only 50 per cent of them also noticed that someone in a gorilla suit brazenly walked onto the court, beat their chest for a few seconds and then wandered off again.

Subsequent smaller studies by Simons and others indicated that inattentional blindness appears to get worse as we get older. If, say, around 40 per cent of people in their early twenties fail to spot the gorilla, between 70 and 90 per cent of those over sixty might miss it. So, Max's grandparents would have even less chance than I did, statistically speaking, of spotting either the red Lego piece or the gorilla.

Given that our brains' already limited attention capacity diminishes with age, it is a perfectly rational decision (albeit an unconscious one) for older people to use the mental strategies that we know from experience will be most efficient. Processing additional information that may or may not be relevant is, on balance, not worth the time and energy. (And energy use is a serious consideration given that the brain consumes approximately 20 per cent of our calorie intake every day. Thinking is extremely hard work!)

But while this adaptation makes sense, it also means that older people are less likely than younger ones to stumble on unexpected solutions or try out new approaches that might be more successful. Not that younger brains are necessarily superior; younger people might not manage to resolve a problem at all, because they get constantly sidetracked by diverging possibilities. It's not that either brain's preferred approach is wrong, more that both are limited. But as our multigenerational Lego triumph shows, these limitations can be worked around if a group can pool and harness their problem-solving abilities.

Family is the first and one of the most formative influences on our intelligence, through both genetic inheritance and learned behaviour. Long before we go to school we learn our first lessons in how to think, approach tasks, communicate and behave towards others by watching our parents, siblings and grandparents. Families of all configurations, from single-parent single-child groups, like Max and mine, to sprawling multigenerational households, create a web of influence that affects how each member tackles problems of all kinds, from building a Lego kit to calming down after a tantrum or resolving a quarrel. The family unit in which we live, especially as children, functions like a laboratory for experiments in the emergence of individual and collective intelligence.

Let's start by looking at what we inherit from our parents in terms of the neural circuitry of our brains, which provides the foundation of our individual intelligence and our aptitude for collective intelligence. We'll then have a look at how interactions with family members, with their distinct brain profiles, can teach us the social and communication skills required for collective intelligence to flourish in a group.

Are We Born Clever – Or Do We Learn Cleverness?

On the whole, families think alike – or at least, children think like their siblings and like their parents. The cartography of any individual's brain is laid down according to a genetic blueprint inherited from the mother and father's DNA. With every unique fusion of sperm and egg a completely new blueprint is created, but we are biologically indebted to our parents in any number of ways, from height and eye colour to the development of our brains. Mechanisms such as recombination and gene mutations have evolved to increase variation, but siblings and their parents share the predispositions associated with particular genes. Our complex behaviours such as impulsivity, resilience and intelligence are all shaped by this genetic inheritance.

With advances in genetics and psychological testing, we can now say that intelligence of the kind that is measured by an IQ test has a high hereditary basis – this is estimated to be around 55 per cent (much higher than the risk of inheriting breast cancer). Most of the genes linked to it dictate how neurons wire up in a baby's brain as it develops in the womb, and how the neurons retain their plasticity (or not) into adulthood.

So cognitive ability is highly heritable – 'clever' parents tend to have 'clever' children. It also seems to be general. There is a lot of data to suggest that if you are good at one subset of the IQ tests you'll probably be good at all of them. It seems that intellectual skills we might think of as quite different (such as mathematics, verbal reasoning and memory) are supported by the same brain foundation.

But if conventional intelligence is 55 per cent heritable, that leaves a massive 45 per cent unaccounted for. Where does the rest of it come from? The short answer is: the knowledge and skills we learn during our lifetimes – everything from learning to read to learning how to take turns speaking so that everyone has a chance to put across their point of view.

Intelligence, like any complex characteristic, is shaped by the interacting effects of hundreds of genes and by multiple environmental factors, which will in turn be influenced by a whole heap of social considerations such as whether everybody can access learning or certain groups are discouraged from it, or unable to afford it.

Intelligence is highly susceptible to both individual effort and collective tweaks. The informal learning that takes place in a family and the official education we receive at school are both crucial, but diet, environment and lifestyle changes have also, to a smaller degree, been shown to make a difference. Given that our families teach us our first lessons in how and what to eat, how to treat other people, how to react to our emotions, how to learn, how to look after ourselves and a host of other important behaviours besides, it's no surprise that intelligence is so shaped by our experience of family.

Learning is integral to the development of intelligence, and the learning that takes place during the first approximately two years of life lays the foundations of the developing brain. A baby learns through every single interaction it has with the world, and, since most of us are cared for by our birth parents, these interactions compound the effect of our genetics, increasing the tendency for families to think alike.

Not that our brains stop developing once we hit our second birthday. They continue to change throughout our lifetimes. Every single idea we absorb and modify, every piece of behaviour we either incorporate into our repertoire or reject, forms and informs our bank of knowledge. Since this is shaped by our genetic predisposition to certain behaviours, our problem-solving abilities and thinking style are the complex result of nature *and* nurture combined. In this way, our family's influence on our intelligence ripples down our lifetimes, via the dance of genetic inheritance and learned behaviour. Before we're even out of childhood, our abilities are in some fundamental sense a collection of skills we've picked up from other people.

Born to Be Empathetic: Babies

One of the most crucial life skills we can learn is to take other people's feelings and needs into consideration. Without an ability to consider and relate to others, our success in life is likely to be limited, in every sphere from friendship to earning a promotion. Empathy is the behaviour that underpins our most fundamental social skills and emotional intelligence, and supports all joined-up thinking.

Our capacity to empathise varies from individual to individual, but most babies are incredibly good at it, beginning to mirror the facial expressions and moods of those around them within days or even hours of being born. Babies reflect what they observe in others without even realising that they, and their feelings, are separate from other people. This is a form of emotional contagion, what cognitive scientists call 'affective empathy', where we feel someone's emotion instinctively after observing their facial expression or other indicators of their mood.

Empathy is an innate skill, already present in the unborn baby. But, like the acquisition of language, it requires exposure and takes practice in order to perfect. When a caregiver behaves empathetically to a child by, for example, comforting them when they cry, they are giving the child experiences that reinforce their pre-wired capacity to empathise. Just as language will be delayed if a child doesn't hear words being spoken to them, so will empathy if a child doesn't receive the message that their feelings are understood and mirrored.

The second form of empathy is cognitive rather than emotional and it takes longer to develop since it depends on the child having developed what's called a 'theory of mind', which is the awareness that other people are separate and different from us and may have different feelings, needs and opinions from our own. 'Cognitive empathy' involves understanding another person's feelings on an intellectual level, taking into consideration their situation and exploring how they might react and what they might need in order to feel better by, for example, 'putting ourself in their shoes'.

These skills are developed throughout childhood and adolescence. If a child of 12–18 months notices that another child is upset, they will probably go to look for their *own* caregiver, whereas a four-year-old is more likely to look for the *other* child's parent. By the time a child is six or seven they are generally better able to regulate their own emotions, which makes them less susceptible to being overwhelmed by emotional contagion and more able to empathise in a wider range of situations.

Caregivers and family members continue to be crucial teachers of empathy as children grow up, but it can be time-consuming work that requires a lot of patience, and a person's capacity to model empathy for the children in their lives has inevitably been shaped by their own experiences. As with every aspect of our behaviour, the legacy of nature and nurture that we inherit from our own families moulds our skilfulness. But even if we didn't receive much empathy ourselves when we were growing up, we can learn to be better at it in adulthood. Shared activity, play of all kinds and making music together are all helpful ways to learn and teach empathy.

Dr Laura Cirelli of the University of Toronto conducted some lovely experiments that showed how exposing a child to singing made them more likely to be helpful to an unfamiliar adult with a problem. Dr Cirelli invited parents of young toddlers to come to a session in which she would interact with the child either by singing a song, reciting its lyrics or remaining silent, while the parents read a story. Then she would encourage the child to watch her complete a simple task such as pinning up some washing on a line. As she was doing this she would drop a peg as if by accident and say, 'Uh-oh'. Children as young as fourteen months old were almost twice as likely to jump in and help her if they had been sung to or had lyrics recited than if she had remained silent previously. In a follow-up study, she found that when a toddler moved and danced with her before the task, they were again much more likely to offer help. Bonds of trust are created by almost any positive interaction but Laura believes that familiar song, the language of lyrics and dancing and clapping along together as we share music, are particularly powerful. This might be due to their capacity to synchronise brain waves between individuals.

Brain synchronicity happens when the seamless portrait of reality that our brain constantly stitches together from the snapshots of information it receives falls into step with somebody else's. It happens implicitly between people, without our awareness, and from a very young age. The result is that people in a group begin to experience reality in the same time frame, and so are more likely to see the world in the same way. Such synchronicity helps to boost learning and consensus-building, and underpins our ability to empathise.

Vicky Leong, associate professor at Nanyang Technological

University, Singapore, is interested in how social understanding emerges and how it links to learning and intelligence. Vicky explained that brain synchronicity has been observed in many different species, including bats and rats, and 'predicts aspects of learning and even social dominance hierarchy. It might be a fundamental, possibly evolutionary conserved mechanism for social coordination . . . which builds cohesion and understanding.' Vicky has done a lot of work with parents and babies in their first year of life, and has observed that brain synchronicity occurs at greater levels when accompanied by direct eye contact and when the parent's mood is positive.

Professor Casey Lew-Williams from Princeton University, who has also done fascinating work on brain synchronicity in parents and babies, discovered that it works both ways. 'We were surprised to find that the infant brain was often "leading" the adult brain by a few seconds, suggesting that babies do not just passively receive input but may guide adults toward the next thing they're going to focus on: which toy to pick up, which words to say.' It seems that there is a dance between individuals in a pair where they take turns to lead the way, even if one is considerably younger. Power dynamics don't necessarily seem to be inevitable.

Many aspects of physiology synchronise between people as they interact; it's not just brain waves, and Vicky thinks they might all be involved in communication. 'Heart rates also synchronise; when we look at things together our gazes synchronise; we share facial expressions. There are many rudimentary forms of social knowledge transmission, which also seem to be available to lots of different species.' Communication emerges from bodies, as much as minds.

Humans also have language, of course, which allows for the transmission of infinitely abstract content. Synchronicity between baby and carer plays a crucial part in language acquisition as the baby imitates the parent, practising the vital communication skills that will support the child as they develop.

It's not just learning how to talk and what words mean in a

literal sense that depends on group learning in the family, though. Recently, scientists have been investigating how we learn to construct and decode meaning itself, as a group project. Understanding meaning, especially nuance and ambiguity, can take years to develop. This process takes place throughout childhood and well into adulthood.

Professor Uri Hasson at Princeton University believes that synchrony is crucial when a group has a conversation and arrives at a shared understanding. Uri studies storytelling in adults, and has observed many different people telling a complicated story to others. The structure of the story remains broadly similar but the exact words used, the phrasing, pacing and background details are completely different every time. Despite this, the high-level meaning is conserved and, what Uri sees when he scans people's brains is brain synchronicity between people, both the different storytellers and the different audiences.

Humans are not alone in communicating meaning and synchronising their behaviour; bees perform waggle dances to others in the hive to direct them towards new sources of nectar, and mice squeak excitedly in anticipation, making vocalisations to each other to communicate emotions. Even plants use their roots to 'listen in' on their neighbours to inform growth.

Humans, however, seem to be unique in dedicating huge amounts of brain power to developing our linguistic skills. There's a theory that competition occurs within a nervous system, meaning that brain territories are fought over on a daily basis, via synaptic plasticity, to retain different skills. Humans invest heavily in protecting our ability to communicate complex and abstract ideas using language.

Vicky says, 'It's inherent in us to want to share our understanding, to co-construct meaning and align our behaviour, and we have multiple ways in which we do this.' Crucially, this capacity for joined-up thinking between people seems to emerge in a baby's brain as they develop their empathy, emotional and communication skills, all of which support each other. We learn the core skills for collective intelligence in our early years and carry on finessing them throughout our childhood.

Vicky has a word of caution, though, on human investment in language and emotional skills. 'The ability to take on board somebody else's perspective, which depends on having a theory of mind, is crucial for empathy but also for manipulation. You can only lie to somebody, for example, if you can grasp that you know something that they don't.' As always, human beings' staggeringly sophisticated cognitive skills, including their ability to share information and emotions, can be put either to positive social use in the service of the group, or to darker and more antisocial use. We'll explore humanity's potential to veer towards the dark side in later chapters.

Natural Creatives: Teens

Teenagers often get a bad press but there are sound scientific explanations for some of the more worrying or irritating behaviours associated with adolescence, such as lying in bed all morning or taking risky decisions. They can be boiled down to saying that the adolescent moment in our brains' development is an incredibly dynamic (and mentally exhausting) period. The prefrontal cortex, for example – the brain region most strongly associated with self-control and judgement – is not fully wired into the rest of the connectome until our mid-twenties. As for the lying in bed, teens typically have different circadian rhythms to older people, not to mention being exhausted by the pace of the change their brains and indeed their whole bodies are going through.

Research by Sarah-Jayne Blakemore, professor of psychology at Cambridge University and author of *Inventing Ourselves: The Secret Life of The Teenage Brain*, shows that the adolescent brain is not a dysfunctional or defective adult brain; it's categorically different. Teenagers are living through a distinct and formative time of life when neural pathways are malleable and passion and creativity run high.

Between puberty and the age of (roughly) twenty-five, neural plasticity is high, as pathways from childhood are reshaped in the light of new experiences. At the same time, the process of 'synaptic pruning' – the removal of less-used pathways – ramps up, while myelin (an insulating layer of protein and fat) wraps itself around some of the nerve cell axons to insulate them. This enables some circuits to develop into a network of highways, increasing the brain's efficiency. These processes come together at high levels during adolescence to create a unique brain stage that offers up great cognitive potential.

Cognitive scientists typically measure two forms of intelligence, both of which are on a steep upward trajectory in the adolescent brain. 'Crystallised intelligence' is your personal repertoire of experience, the breadth of knowledge that you have about the world. 'Fluid intelligence' is your capacity to apply that body of acquired knowledge in creative, flexible and reasoned ways. Teenagers obviously have less of the former but are driven to get out there and seek new experiences precisely in order to increase it. Meanwhile, they are getting really good at fluid intelligence and processing speed. The teenage brain is open to new knowledge and relatively free of assumptions about the world. It is particularly adept at lateral thinking, highly creative and fizzing with intellectual possibility.

Creativity hasn't been much studied as a distinct component of intelligence until recently. Alison Gopnik, professor of psychology at Berkeley, University of California, is one of the people changing that, and a lot of her research is focused on children and adolescents. She has looked at the way that children play and teenagers use exploratory, creative and learning behaviours, as well as the trade-off that occurs later on, as we shift into adulthood. Adults usually focus less on exploring and more on exploiting the knowledge they have already built up.

This work ties in neatly with Sarah-Jayne's research into the prolonged prefrontal changes that you see in the human child and teenage brain, which takes much longer to develop, relative to lifespan, than any other species'. The prefrontal brain regions directly behind the forehead are associated with executive functioning and decision-making: the sorts of skills that make it easier and more rewarding to exploit knowledge rather than explore the world and exercise creativity. Sarah-Jayne maintains that it's this extended phase of brain development that has allowed our species to develop its phenomenal creativity, and that it is crucial for language and social learning.

Delays in connecting up that executive region of the brain can have drawbacks and are linked to diagnoses of ADHD, but there does seem to be a definite advantage to creativity if prefrontal brain regions are less active. Dr Hao Zhang from the School of Psychology, Southwest University, China carried out a lovely study into a very simple way to dial down its activity. He found that creativity can be improved at any age simply by breaking off from a task and going for a walk, the caveat being that you can't follow a prescribed route, you just have to wander around with no goal in mind. Diverting activity from the decision-making prefrontal cortex to the muscle-moving motor cortex seems to free up mental energy for lateral thinking. A number of similar studies indicate that engaging in simple tasks that allow the mind to wander, such as washing the dishes or going for a gentle jog, helps your alpha and gamma electrical oscillation waves to dominate so that when you return to a task, your mind will be clearer and your creativity higher.

Acknowledging the raft of changes going on in the adolescent brain might mean we alter the way life is structured, to accommodate them. Sarah-Jayne advocates a later start to the school day, for example, and abolishing exams at the age of sixteen since they take place in the middle of a period of intense change in the structure of the brain. And on the flip-side, perhaps we can do more to harness the enormous creative brain power of teenagers. Globally, around 15 per cent of our population are teens. Imagine what we could do if they were invited onto decision-making committees and steering groups. Perhaps it would be the boost to our species' collective intelligence that we need.

Wisdom and Expertise: Older People

Babies may be champion learners and teens may be cognitive wizards, but it's not all bad news for us older people. This graph shows what typically happens to crystallised and fluid intelligence over the course of a lifetime.

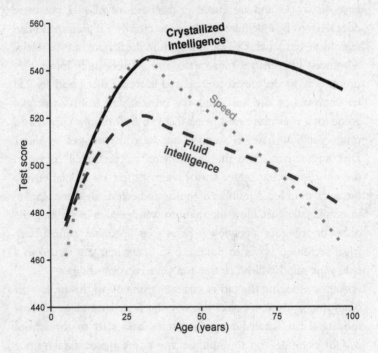

Change in cognitive abilities with age, adapted with permission from Tucker-Drob, 2009 and featured in Stuart Richie's book Intelligence Matters, *published by John Murray Learning.*

Take a look at the curve for crystallised intelligence. We tend to acquire more and more of this until we have a bank of what we might call wisdom, or expertise. All of us are experts in some areas of life, and the older we get, the more expert we become. Time spent building up deep skills in a specific area leads to changes in the brain as masses of new connections form between nerve cells, enabling efficient communication routes to form.

If you take images of the motor cortex region of the brain as somebody learns the piano, say, you can see it bulking up, just as our muscles bulk up at the gym. (The motor cortex is the brain region involved in coordinating and directing the compli-cated hand movements required for fluent playing.) Similarly, a well-known series of experiments conducted by Professor Eleanor Maguire of University College London showed that London cab drivers have enlarged areas in their hippocampi. The hippocampus is key to the workings of memory, and cabbies famously train to know their way from any point in Greater London to any other, a body of expertise referred to as The Knowledge.

Whether it's playing the piano or driving a London cab, after an intense period of learning the brain no longer needs that bulk of neural pathways and can afford to prune them back to hyper-efficient mega-pathways. Recently this mesmerising sculpting process has been revealed. Mriganka Sur, professor of neuroscience at Massachusetts Institute of Technology, established that once a connection between neurons has been strengthened to a certain point, a genetic switch is triggered that causes neighbouring connections to dissolve. In this way the brain optimises its circuits and settles into efficiency. Individuals can now automatically use the skill they have acquired, circuits are refined to their key pathways and the brain's volume reduces once again. This elegant mechanism accounts for the fact that we can all learn so many different skills over our lifetimes. We will not run out of room in our skulls to accommodate our bulky brains because thankfully there is a way to retain expertise within the neural maze that tracks its development.

So as far as crystallised intelligence goes, we are on the upward trajectory throughout our lives. On the other hand – and there's really no way to sugar-coat this – our fluid intelligence starts to decline from our thirties onwards. (Plot your age on the second curve of the graph and feel the fear.) This is connected to that strategy we saw earlier for coping with declining flexibility and memory capacities, and explains why there's typically less innovation and creativity among older people.

If you're already past the peak of the fluid intelligence curve, console yourself with the thought that there is robust data to show that unless you are unfortunate enough to succumb to dementia, your fluid intelligence will not completely desert you. In fact, a high score on an IQ test (a measure of fluid intelligence) in adolescence is a strong predictor for having a similarly high score in later life. Admittedly you might score lower aged eighty than you did when you were twenty but if you scored high at twenty in comparison with your peers, you'll still score high at eighty.

And family life is the perfect brain tonic for older people, who will benefit as individuals and contribute powerfully to collective intelligence by bringing their expertise and wisdom to a multigenerational group. The mutual cognitive benefits of interactions between generations are profound. Studies have confirmed that the act of exploring and interacting with others helps keep the mind and body healthy. Older people's brains stay active by playing and exploring. We're probably all familiar with the idea that new challenges can keep our brains young, and that's absolutely true on the level of neurons and synapses. The creative, novelty-seeking, hyper-flexible cognitive capacities of a child or an adolescent can push the older brain to challenge itself, and the child's brain benefits from contact with the wisdom and expertise built up over a lifetime of experience.

The emotional bond between generations and the empathy

it's built on also seem to support good brain health into later life. Scientists recently scanned grandmothers' brains while they looked at photos of their grandchildren and saw areas of the brain associated with emotional empathy light up. The grandmothers related to their grandchildren emotionally in a much more synchronised way than they did with others, even their own children. This tendency appears to bring a benefit to the older generation, with grandparents who spend more time with their grandchildren showing less cognitive decline on average.

Family life provides a living, breathing Petri dish for collective intelligence, showing us how all these differently structured brains feed their specialisms into a group's repertoire of problem-solving skills.

Men and Women – Is There a Cognitive Difference?

We tend to categorise our species by sex (as well as gender): human reproduction being what it currently is means that two distinctly different components – male and female sex cells – are necessary to create another person. Our understanding of sexual and gender identity and how they operate within family life has evolved alongside developments in reproductive science. Families now come in all shapes, sizes and identities but since most of them include some male people and some female, it's relevant to question whether they solve problems in different ways, or contribute different skills.

In 2019, I had an opportunity to observe how biological sex can play out in a family's ability to solve problems when I was the science presenter on a BBC television series called *The Family Brain Games*. It was a knockout competition that pitted families from a wide variety of backgrounds against each other. Some families had members from three generations, some from two. A laboratory full of ingenious team games was created and they all had to complete a series of puzzles, against the clock, and all, rather dauntingly, in front of a film crew. Perhaps unsurprisingly, it was the children of the programme who ruled the roost, especially for the creative-thinking puzzles.

What did the winning team do to ensure success? Well, the way they related to one another and communicated was impressive. They excelled at keeping each other calm by identifying how each one was feeling and synching their emotional states to the most relaxed, through prayer and song. They trusted one another, listened to one another in turn and assigned roles within the group according to ability rather than status.

At the outset of the competition I was asked to predict which team would win. Somewhat to my relief, I turned out to be right. My reason for picking this particular group was the high number of female participants (two young daughters and their mum, the dad being the only male). But before I'm accused of gender bias, my prediction was based on some scrupulous research carried out by psychologist Anita Woolley of Carnegie Mellon University and her colleagues.

They conducted two studies that examined the individual intelligence of 699 people. Then they put them to work in groups of between two and five people and each team was asked to complete several tasks, including visual puzzles, brainstorming, decision-making and a complex problem. Woolley's team found 'evidence of a general collective intelligence factor that explains a group's performance on a wide variety of tasks'. The strongest predictor for how well a team would do was . . . yes, the ratio of females: the higher the better. Woolley suggested that this might be linked to a female tendency (possibly cultural) towards higher social sensitivity and for less interrupting and more listening.

The issue of whether men and women have structurally different brains or are socialised to have different strengths and preferences is a hot potato, to put it mildly. Scientists have been arguing about it for years. Some studies suggest that there is a small but appreciable difference, perhaps due to the influence of sex hormones on the developing brain while the baby is still in the womb. Others say any differentiation is so tiny and with so much overlap between individuals that it is unscientific to assume categorisation based on biological sex.

Increasingly scientists are converging on the concept of a brain continuum related to biological sex. A recent study by Professor Barbara Sahakian at Cambridge University and colleagues at Fudan University, Shanghai supports this. The study assessed connectivity markers (the way that different brain regions connect up and shuffle information between each other) to characterise brains as either 'male', 'female' or 'androgynous' in terms of their neural circuitry. It used three very large data sets from different cultures: 9,620 participants in total (4,495 men and 5,125 women). The results found that approximately 25 per cent of brains were identified as 'male', 25 per cent as 'female' but the majority – 50 per cent – were distributed across the 'androgynous' section in the middle of the graph on which they were plotted.

Cognitive 'androgyny' had clear benefits to the individual, irrespective of their biological sex. These people had fewer mental health symptoms such as depression and anxiety compared with those at the extremes of the spectrum. The 'androgynous' brain was also linked to better cognitive flexibility (the ability to shift between different tasks or thoughts), higher social competence and more creativity. The study's authors concluded that 'for optimal performance in school, work and for better wellbeing throughout life, we need to avoid extreme stereotypes and offer children well-balanced opportunities as they grow up'. Hard to argue with that.

Men and women may or may not have neurologically different brains but they sometimes behave in very different ways, though it is generally accepted that a huge amount of this is learned. Often the learning takes place unconsciously through observation, conditioning and repetition. Fewer girls these days are told that science and engineering are not suitable subjects for them, for example, but girls might still absorb these beliefs because they continue to circulate in our society. The same thing holds for boys and men, many of whom learn that they don't need to prioritise emotional self-awareness or cultivate empathy. They might even absorb the lesson that such things cannot be learned. This belief can have serious negative consequences for them as individuals. It does boys a disservice and means we are not making the most of their potential contributions to our species' future collective intelligence.

As things stand, adult women are typically more skilled than men at turn-taking, active listening, empathy and all the other pro-social behaviours that support effective group work. And this translates to better outcomes for groups: more women basically means faster problem-solving.

Given that sex and gender are among the most fundamental sorting categories we use to think about the world, it's no wonder that they are so powerful or that they're subject to a great deal of unconscious bias. But bias can be challenged by bringing our unconscious beliefs to our conscious attention. In this way we learn how to draw on more aspects of our own intelligence, by tapping into different skillsets and abilities. We can create new patterns of thinking and learn new behaviours.

So what were the learned behaviours that the successful team deployed in the *Family Brain Games* challenges? Aside from the genetic inheritance they were born with, aside from the questions of age-related and biological sex-related diversity that we've looked at, what did the winning family *learn* how to do that supported their success?

Share Power and Responsibility

First off, the most successful families all adopted a non-hierarchical approach that seemed to be very much embedded in how they related to one another. Everyone participated, from the youngest children to the oldest grandparent. They formed intergenerational alliances. People with specific knowledge were asked to share their opinion with the others. They worked alone when appropriate and then came back to the group to discuss. They moved to resolve emotional upset quickly. They looked each other in the eye when they spoke and they didn't interrupt; everyone was listened to with full attention.

These behaviours may sound obvious, perhaps a simple matter of good manners, but manners alone will not cut it if we want to see authentic collective intelligence. Relying on norms of politeness leaves us at risk of unconscious bias because not every individual is necessarily seen as deserving of the same respect. We all have to learn how to listen, ask for advice, accept a criticism and change our minds. These are cognitive habits that cannot be turned on at the switch of a button, even with the best will in the world. If they are embedded in our repertoire of behaviour when we're young, through our family interactions, they become part of our mental furniture, laying the foundations for the social and emotional intelligence we will need throughout our lives.

Boost Mood and Build Consensus

Besides the learned communication skills and awareness of power and status that the family groups brought to the tasks, they also carried out certain activities to control their moods and shape their group dynamics. The families that made it to the final rounds sang, laughed, meditated and prayed together as they waited for their turn. These interactions all have a positive effect on individual and group performance. Laughter is a mood-boosting tonic for anxiety since it is accompanied by the release of endorphins such as serotonin, which are linked to feelings of wellbeing. Any kind of shared activity reduces the build-up of the stress chemical cortisol, which can impair neuronal plasticity and lead to frightened brain-freeze.

Such activities also enhance electrical oscillations, helping to utilise more of each individual's brain power by boosting the rate of connectivity across the whole brain. And singing or praying together can synchronise people's brain waves so that they start to literally see the world at the same timestamp, which, as we know, makes it easier to build a consensus among individuals.

Meditate for Greater Lateral Thinking

Families used group meditation sessions backstage as a way to help them set aside the stresses of competing on the show. Interestingly, the calm experienced by one member of the family seemed to transmit, during a guided meditation, to others. There was a contagious effect thanks to the synchronising of brain waves brought about by the shared activity. But beyond this calming effect, meditation also seemed to support the families to think creatively. Mindfulness widens our range of perception and allows us to think outside the box.

Studies on Buddhist monks, professional gold star meditators if you like, have shown that over the course of extended meditation, different brain waves that are associated with different cognitive states are enhanced. Gamma waves are the brain's fastest speed of electrical activity and are thought to engage disparate brain regions, drawing together the whole connectome. Theta waves, which are slower and more rhythmical, also increase. These are associated with enhanced learning and memory. Thirdly, there is an increase in alpha waves, indicating that the brain is taking a rest from intentional, goal-oriented tasks. The cycling between these different speeds of activity supports the brain to attain moments of insight and then incorporate them into its stock of knowledge. This in turn supports creative problem-solving and deep learning, as well as relaxation.

Build a Shared Memory Bank

Family groups were able to support each other and get better results than they could individually, by creating a pooled memory store. Short-term memory suffers when we are stressed or tired, as well as diminishing with age, so it makes sense that our brains have evolved to be efficient at creating what scientists call 'implicit' or 'transactive' memory.

The concept of a shared memory bank co-created by a group of people who know each other well was first explored by Paula Raymond and Daniel Wegner at New York University. They set up a memory test with fifty-nine couples: half of whom were asked to work together while the other half were split up and assigned a partner who was a stranger. All pairs were presented with sixty-four random statements on subjects unconnected to their personal history over five minutes, then given a distraction task for a few minutes before being asked to try to remember as many of the statements as possible.

The result? The pairs who knew each other had a 14 per cent memory boost compared to those who had just met for the first time. It seems that the successful couples knew each other well enough to predict which items their partners would best remember, without even being consciously aware that they were doing so. They knew instinctively, or rather they knew *subconsciously*, to focus their own mental energies on the remaining statements. The strangers, unfamiliar with each other, had no shared memory bank to deploy and had to double up on the work of remembering.

There is a poignant downside to this that affects couples who end up divided, either by death or divorce. The loss of transactive memory can feel like losing a part of one's own mind. In clinical neuroscientific terms, the loss we feel at the end of a relationship might be partly due to the loss of our transactive memory; our memory partner has effectively crashed, reducing our own cognitive capacity.

It doesn't take a lifetime to achieve this memory-merging, though; the effects are measurable after about three months. Given this relatively short time frame, and since the booster effect of paired cognitive capacity is so powerful, it might make sense to try to build a shared memory bank with work colleagues (while obviously trying to avoid any deep grief at the end of a project!).

Laying the Foundations for Collective Intelligence

For most of us family life is our introduction to group living, where we learn to regulate our emotions, empathise, communicate with others and tolerate the idea that we are not always right about things. Not all family environments are supportive or even safe, sadly, but most of them provide a relatively stable introduction to getting along with and working alongside other individuals. Since family members share genetic profiles and environmental experiences, they are typically able to experiment with communication and learning in a sympathetic space. There will always be differences and conflicts within any family but this can teach us the skills we need later in life as we venture further afield and encounter people with very different perspectives.

Families don't necessarily share genetics or relate to one another in the traditional multigenerational format, of course. There are many blended, adopted and fostered families flourishing in any number of different ways. These more cognitively and socially diverse family units offer their members exposure to a greater range of perspectives and may foster more innovative thinking styles than the classic nuclear family of two parents and two children.

Max and I benefited from precisely this sort of self-created family during lockdown in Queensland. When I realised that the situation was likely to last a long time, I set about creating a network of friends, embedding ourselves within the local community and building us a new type of family. Thankfully (and admittedly I might be biased here), my son is exceptionally sociable and fun-loving – in dog terms, I'd compare him to a beautiful golden retriever puppy – which was a definite help when it came to making friends. We joined the community garden, where we planted trees, grew veggies and met a whole network of people. My son now calls one fabulous couple his adopted grandparents. Ros and John were missing contact with younger generations just as we were missing older people, because their grandson lived behind the closed borders of a different state. We formed a pandemic adoptive family together and also joined forces with another family when we rented the spare granny flat in their garden. They generously invited us on holiday with them, and to join their grandparents and sisters, cousins and aunties for a massive Christmas Day lunch.

The strict border closures were tough on many families but they kept the virus out and prevented lengthy lockdown measures. Along with the exceptionally friendly and kind community we found, this meant that Max and I were able to avoid the worst of the mental fog, loss of short-term memory and general malaise that scientists have associated with pandemic living. We thrived during a time when we could have been lonely, and I could not be more grateful for our unexpected stay.

It is sometimes said that we are living in an age of chronic loneliness, and the statistics certainly show that for some people, social isolation is a horrible reality. But at the same time we have more freedom than ever to decide who to live with and how. We can choose to blur the edges between family and friends. Our birth family might lay the foundations of collective intelligence but it is not separate from the wider community. If we are lucky, it can be the model that teaches us how to contribute to networks of cognitive richness and support, showing us that this collective can help instil resilience, buffering against the difficulties that life will occasionally throw at us. Most importantly, this family – in whatever form or shape it takes – can help to make life and learning fun.

EXERCISE: Practise the Art of Listening

Set aside time with your family to sit with each other, to talk and listen. Each person has five minutes in which they can speak, without any interruption, about any topic they choose. Family members can either practise their direct eye gaze, looking at each other as they talk if they feel comfortable, or look away if they prefer. The other members get to practise active listening, giving the person who is talking their full attention without jumping in with any comments or questions. This can feel a little strange at first as we're so used to the back and forth of conversation but it can be immensely liberating and empowering, and builds a deep connection between individuals.

Here is a list of questions you could use to get started:

- If you could wake up tomorrow having gained a superpower or new ability, what would it be?
- Would you like to be famous? In what way?
- What would constitute a perfect day for you?
- For what in your life do you feel most grateful?
- If you could change anything about the way you were raised, what would it be?
- Tell your life story in as much detail as possible.
- Is there something that you've dreamed of doing for a long time? Why haven't you done it?
- What is your favourite memory?

With younger family members you can play a game. Place a number of objects such as a toy horse, car or teddy bear in front of you, and look at them in turn. Then settle your gaze on one and ask the child to tell you what you are looking at. Once you've selected an object, take it in turns to create imaginary scenarios with it; so for example, you might tell a story about the horse, what it might be feeling, where you might go on adventures together, what it might have been doing that day.

CHAPTER FOUR

Intelligence at Work: Recruiting the Right Team

One fresh autumnal morning in a pre-Covid world, a gaggle of academics gathered in the courtyard of Magdalene College, Cambridge. I had asked some colleagues if they would be willing to participate in a test of our collective intelligence and they had kindly agreed to take part in an escape-room challenge. These adventure games have become hugely popular in recent years. Originating in Japan, the concept spread across Asia before hopping over to America and Europe. By the start of 2020 there were over 10,000 escape rooms worldwide.

The premise is simple: people pay a fair amount of money to be locked up, usually in a dark and gloomy place, and the team has to work together to solve a series of cryptic clues in order to access a key and escape. All the while, the clock is ticking. Frankly it's my idea of claustrophobic hell but, when I discovered there was an escape-room company that would bring their portable 'room' to us, it seemed too good a research opportunity to pass up. I plucked up my courage and recruited some more-or-less willing victims.

That's how we all came to be waiting for Phil, the manager of Cryptic Experiences, to arrive with his kit on that breezy morning as the sense of apprehension mounted. Our motley crew consisted of the college chaplain, two mathematicians, a historian and me. Would we be able to work together to solve ten mind-boggling puzzles in time to secure our release? Or would we be trapped in the escape room for brain-aching eternity?

I won't keep you in suspense: we acquitted ourselves reasonably well and managed to work in efficient harmony to secure our release. Which was a relief. We certainly didn't set any records for a quick exit, though. Phil told us that he'd had a team of actuaries in the week before and they'd completed the task a full five minutes faster than us.

Our cognitive adventure required us to deploy skills that are integral to collective intelligence in any situation but are particularly important once the group moves beyond the bonds of family and kinship – in the workplace, for example. Most of us are paid to complete tasks and fix problems of one kind or another, and most of us work with other people to whom we are probably not related and with whom we haven't chosen to associate, in groups of various different sizes. The central question of this chapter is, how does collective intelligence function in these circumstances?

I wasn't sure what to think about our chances in the escape room. Did we have enough diversity? We all had quite clearly delineated areas of expertise spanning many different areas, which was a plus; but expertise in itself can be a predictor of cognitive blindness. We were two women and three men; we were similar ages but had grown up in various countries across the world. There was no hierarchical power structure and everyone was respectful and curious. Would this be enough?

And what about our ability to use intelligent strategies to boost our joined-up thinking? We would have to deploy creative thinking as well as draw on prior knowledge; so we would need to inhibit assumptions and accept ambiguity in order to allow information to connect in new ways. We would then have to trust that everyone had something to contribute, would listen attentively and take on board each others' perspectives. Only in this way would we be able to pool our breadth of expertise and thinking styles.

Fortunately, and much to our relief, we were like a well-oiled machine, reading out the questions clearly, sharing information and then brainstorming possible solutions. We instinctively did a fair amount of direct eye-gazing while mulling over the puzzles, which helped to synchronise our brain waves and support us to reach a consensus. Every person was listened to and this respect and our confidence that we would be heard proved absolutely necessary, since each person provided at least one key idea that helped to solve a clue.

In this chapter we'll be looking at why cognitive and social diversity among teams is the foundation for success on any project, whether it's based in the office, out in the community or in an escape room. As we will discover, robust scientific studies have shown that the recipe for group success is to bring together people with different experiences who are confident in their abilities, have fairly equal levels of cognitive competence and the ability to communicate freely.

Why are these factors so important? Well, without a diverse range of viewpoints there's a risk of insufficient information or a missing skillset. With too much narrowing of the group's perspective, blind spots and bias are amplified. Consensus thinking can solidify into an echo chamber. This is true even if the individuals are experts, brilliant innovators, great leaders or proven project managers. We'll see that experts and creatives, for example, are both necessary but also insufficient on their own, and why introverts and extroverts, team players and leaders are all welcome and all valuable. We'll explore how we can put together a strong team and help them work together to make use of all the brain power on offer.

Setting up Our Group for Success

Once we move beyond family and kinship groups where participants are bonded by genetics, the size of the group and its attitude to the difference and similarity of its members become crucial. These factors work either to increase or decrease the likelihood of intelligent behaviour emerging in mixed groups of all kinds, from friends to housing cooperatives.

Let's start with group size. Does that have an impact on a group's ability to communicate effectively, take decisions, bond, empathise and get things done? Does decision-making inevitably become too complex as we form a larger collective? Anybody who has ever found themselves on a group holiday trying to agree which restaurant to go to for dinner might well think so.

Human beings are creatures of staggering ability when it comes to collective action, as Robin Dunbar at Oxford University has shown. He has analysed his 'social-brain hypothesis' on a species-by-species level, and corroborated that the relative size of the ventromedial and orbital prefrontal cortex regions (so just above and behind the eyebrows) correlates with the size of social groups. These regions are associated with executive reasoning in mammals, and particularly in primates. Their size peaks with us humans and, according to Dunbar, enables us to maintain stable social relationships with, on average, around 150 people. Beyond this approximate size of community more formal rules are required to govern social dynamics. As we've already seen though, humans have proved adept at coming up with those rules and systems, from trade guilds and courts of law to Trustpilot.

Long before I met Professor Dunbar I had spent a decade living in an environment where being part of a team and working as a collective was essential for the basics of life. I lived off-grid in a houseboat on the River Cam in Cambridge, UK. A tight-knit community was utterly essential: this was before social media or widespread Wi-Fi and before efficient solar panels and battery devices were commonplace. We had to generate our own electricity, bring fuel onboard for heating and share a single water point for the whole community. There was no boatyard for miles and it took weeks to get somebody out to look at electrical or gas problems when they (inevitably) arose.

In times of emergency, all hands were on deck to help out. We all developed problem-solving skills, resilience and resourcefulness, and passed on our experience and knowledge to our neighbours. We relied on collective wisdom passed down from community elders and on peer-to-peer sharing. Our collective endeavours built bonds of trust and friendship that in turn reinforced our ability to tackle problems. The boating community consisted of, you've guessed it, just under 150 people – it's capped by city restrictions on boat numbers.

While I was living on the boat I was also working as a Fellow of Magdalene College, just a few hundred yards upstream of my mooring. And, you've guessed it again, the size of Magdalene College Fellowship is not far off the magic 150, currently standing at 125.

This size of group seems to facilitate the transmission of ideas. It's big enough to encompass diversity of expertise and perspective so that it avoids becoming an echo chamber, but small enough for discussions to be manageable. All over the world, groups settle around the same number. Silicon Valley, for example, certainly houses more than 150 people overall, but small hubs for exchanging ideas and information have always been integral to its success. They helped Silicon Valley to overtake its competitor tech hub, Boston Route 128, where hierarchical and isolated companies that were protective of their ideas failed to thrive. Silicon Valley success stories since then are legion. The Apple Corporation, to take one iconic example, was born out of a Homebrew Computer Club meeting held in a garage in 1975.

So if humans are so good at working as a group, right up to a collective of 150, how do we account for those squabbles over which restaurant to choose?

The obvious answer is that even if our group is much smaller, perhaps only eight or ten, we might know, believe or want different things. Wherever there is difference of information held, opinion or desired outcome, that presents an opportunity for a great solution to emerge, but also an obstacle to its achievement.

That's why diversity on its own will never be enough to guarantee joined-up thinking. Diversity without empathy, clear communication and awareness of self and others – social intelligence, in other words – is liable to lead to conflict and chaos.

Human beings are great at finding ways to limit diversity for precisely this reason: because it's cognitively easier to limit the amount of difference in the people we interact with than it is to open our minds to doing the hard work of relating to somebody who is very different from us. Much of this behaviour is damaging and unjust and limits our potential for collective intelligence. We will be talking about unconscious bias in relation to all these worries later in the chapter. But it's also worth acknowledging that some of it is benign, especially if that acknowledgement helps us to accept that these choices over who to listen to and talk to are going on at a subconscious level for *all of us,* all of the time. Luckily we can, with intention, change them.

Take friendship groups. Unlike families, who are connected by genetic similarity, friends tend to be connected by bonds of social similarity. We have something in common, whether that's our background, our school, a hobby, a cause or a combination of many factors that drive us to pick someone out. This selection process is sometimes conscious but mostly unconscious. It's yet another of the brain's energy-saving strategies. We choose friends on this basis so that we don't have to waste cognitive resources on figuring them out, only to discover that they're not our kind of person after all.

This sounds cynical until you remember that firstly, we're not conscious of it most of the time, and secondly, friendship is – like intelligence itself – a survival strategy. Homo sapiens needs allies to see us through conflict, illness and hard times. A human being with no clan is still, as we have seen, extremely vulnerable to mental and physical health problems. So friendship is a high-stakes game that's worth playing as intelligently as possible.

Professor Dunbar suggests that, 'We each have the friendship equivalent of a supermarket barcode that advertises certain key characteristics. It is activated primarily by our use of language, which makes it "scannable" by others when we interact with them. The more characteristics we share with another individual, the higher the chance of us hitting the loyalty-card jackpot and forming a close friendship.'

But as usual, our flawed brains are liable to make mistakes as they cut corners. Our mental laziness might save energy but it's also a block on our learning, creativity and innovation. If we want to counteract our brains' preference for safe choices, we must bring conscious awareness to the process of choosing friends and of deciding who to pay attention to in the first place. We will need to deliberately seek out different perspectives.

Our flawed social intelligence, so agile and yet so prone to bias, helps to explain why diversity is essential for intelligent *solutions* to emerge but insufficient on its own for intelligent *group behaviour*. Only when the group has put in place structures to counteract the damaging effects of unconscious bias can the full power of diversity be unleashed. Then, the individual's cognitive world expands and the species' mental evolution can accelerate.

Of course, we are all biased, just as we are also all hard-wired to find new ideas rewarding and stimulating. This paradox is the context for everything else we talk about in this chapter, and the next one. Any arena where we are trying to work together towards some purpose will generate more joined-up thinking if it contains a cognitively rich tapestry of people bringing together their various genetic heritages, behavioural habits, foibles and idiosyncrasies. But without awareness of the purpose of diversity, and appreciation of every individual, the group will struggle to perform intelligently.

Diversity Is the Key to Intelligent Groups

Why and how does having a diverse team make such a difference to outputs and performance? Firstly, there's the intuitively obvious point that if you have a greater range of information and perspectives available, that has to be a good thing. Chad Sparber, professor of economics at Colgate University, New York, conducted some stellar analysis to calculate how increasing racial diversity in the American blue-collar workforce by just one standard deviation point results in increased productivity of more than 25 per cent across the board.

His data also suggests that while immigrants make up only 13 per cent of the population of the United States, they have accounted for about a third of the US's patents since 2000. Then there's the data demonstrating how a 1 per cent rise in employment of non-US nationals in science and technology sectors increases the wages of college-educated native-born people by between 5.6 and 9.3 per cent. Recruiting from a global pool of applicants and having a culturally diverse team pays off, both longer-term for innovation (as measured by patents) and in the shorter term, in higher productivity and wages.

Beyond this diversity of knowledge and perspective – of content, if you like – there's also the diversity of thinking and decision-making styles embodied in the brains of all those different people. They have not only different knowledge but also different ways of working with that knowledge to come up with new and creative solutions. Given what we now know about the extent of the brain's responsiveness to context and its capacity to generate new cells and rewire the neural pathways, it makes sense that different lived experiences will provide different strengths. This multiplicity of thought processes is particularly important when it comes to innovation, where different perspectives can contribute to very different ways of looking at and solving problems.

The ability to think laterally is perhaps the defining characteristic of intellectual work in the twenty-first century. It's the difference between knowing a fact and imagining what you can creatively and usefully do with that fact. Individuals with different ideas are intrinsically valuable in this context. When a group has worked out how to pool its variety of knowledge and thinking styles and work on them together, the conditions are ripe for *group* lateral thinking. And this is where having a diversity not just of ideas but of decision-makers and creators to work with them really fires up collective intelligence so that a team can operate at a grander scale.

In the field of scientific innovation, diverse teams have proved themselves to be outstandingly successful. Brian Uzzi, professor of leadership at the Kellogg School of Management, Northwestern University (and his team!) scoured a whopping 2.1 million patents and almost 20 million publications over five decades, looking for the patterns underpinning great success. Beyond the general conclusion that teams beat individuals consistently, they found that teams of people combining atypical subjects across different disciplines were far more innovative than more homogenous teams. Strikingly, this advantage has been increasing over time, presumably in line with the world of academic science becoming increasingly specialised. Uzzi concluded his study by saying that, 'The process of knowledge creation has fundamentally changed.' It is no longer enough to be an expert. Over-reliance on individual ability is inefficient. We are most effective at solving problems when experts from different fields work together.

Diversity is also crucial for the avoidance of bias, narrow perspective and blind spots. We've seen that any individual's brain is limited in its perception and its understanding of reality. That's one of the main reasons that we're social animals and hard-wired for collective intelligence. This has profound implications for the ability of any group to function in an intelligent way.

Jane Goodall is one of the most famous and eminent scientists at work today, and one of a number of pioneering female primatologists that also includes Dian Fossey and Biruté Galdikas. Studying the behaviours of our nearest non-human relatives can sometimes lead to deterministic pronouncements about human behaviour. Primatology is particularly vulnerable to bias and blind spots, in other words, as Goodall's work demonstrated.

Male research into the behaviour of chimpanzees had for years been working on the assumption that the chimpanzee social structure depended on male rank order, with females passively fitting in. It wasn't until Goodall arrived, and focused her study on their behaviour in their native wild environments, that she noticed that females solicit sexual activity from a wide variety of males, not just the higher-status ones. She set about investigating this behaviour and discovered that females were also systematically killing off younger chimps in the troop in order to maintain their dominance, even going as far as cannibalism. Goodall's male colleagues had presumably also seen the female chimps behaving in these ways. They had simply ignored it as irrelevant because it didn't fit with the accepted theory. The male scientists had a blind spot.

As in primatology, so in any field of endeavour. Diversity of all kinds is important to ensure that the right details are being noticed and followed up and all the possible questions are being asked. Without it, our blind spots will massively reduce our collective intelligence.

How Diverse Teams Work Best: Balancing the Skills

The case for recruiting people from a range of social backgrounds into our teams is absolutely rock solid on any metric, from driving innovative solutions to improved productivity. Beyond the general argument for a socially diverse team, how do specific kinds of *cognitive* diversity align with some of the different skill-sets that are valued at work?

Let's start with experts. Every project will benefit from having input from someone with experience and specialist knowledge. There are many situations that require expertise to arrive at a solution, from removing a brain tumour to changing the tyres on a Formula One racing car. The value of experts is to a certain extent self-evident – you wouldn't want your brain surgery performed by a car mechanic, however skilled – but expertise is not without its drawbacks and increasingly its benefits are most powerful in a team context.

As we saw in the previous chapter when we looked at the effect of ageing on the brain, any accumulation of knowledge is both a resource and a limitation. Generally speaking, the more expert we are, the more we rest on our laurels and fail to investigate new information. This explains why, as science journalist and author David Robson points out in his book *The Intelligence Trap*, a higher level of education is not always synonymous with astute thinking. Robson rather pointedly states, 'Intelligent and educated people are *less likely* to learn from their mistakes for instance, or take advice from others.'

Higher education, or simply having lived a long life and gathered a lot of knowledge, can result in blind spots, *less* cognitive reflection and making more assumptions that lead to the wrong answer. This is obviously a problem for the individual but it can also be a significant barrier to collective intelligence if the expert person not only rejects advice from others but doesn't even realise they need it in the first place.

Robson provides a lovely example of this in the form of a logic problem that was printed in the *New Scientist* magazine as a brain-teaser. See if you can get it. (The answer is in the chapter references at the end of the book.)

Question: Jack is looking at Anne but Anne is looking at George. Jack is married but George is not. Is a married person looking at an unmarried person?

Answer: Yes, No, Cannot Be Determined

Most of the readers who submitted an answer got it wrong. Even worse, an unprecedented number of them wrote to the magazine claiming the printed answer was a mistake. Why? The readership of *New Scientist* is highly educated and skews towards the over-fifties. In other words, they're experts. Their keenness to reach an answer quickly, combined with their overconfidence in their capabilities, meant they were unable to think logically through the problem.

As individuals, we all have to carry out a balancing act every time we take a decision or attempt a task, trying to harness our expertise while avoiding the narrowing of perspective and blind spots that come along with it. Sometimes we manage it, sometimes we don't; but it's much easier to do if we're part of a team. Collectively, it seems, the wisdom accumulates and errors nullify themselves, an idea we'll return to when we look at the concept of the wisdom of the crowd and its possible links to our intuitive intelligence.

One counterbalance to the potential narrowness of expertise is the originality that comes from lateral thinking – in other words, not just having the knowledge but being flexible and innovative in how we apply it. Creativity of this kind is highly valued in many businesses and in our culture in general, and as we've seen, it's coming under more and more scrutiny from scientists. So can it be nurtured across a group of people as well as by individuals?

Claire Stevenson, assistant professor of psychological methods at the University of Amsterdam, worked with Eveline Crone, professor of cognitive neuroscience and developmental psychology at Leiden University, to investigate these questions. Their studies combined mathematical modelling with psychological testing and focused on adolescence because, as we know, the younger brain's hyper-plasticity inclines it to higher creativity.

What sort of lab experiment can measure the famously unpre-dictable and potentially wild creative process? Well, Stevenson and Crone used the Alternative Uses Test, in which people are asked to come up with creative uses for an everyday object. What could you do with a rolling pin, for example? Sure, you could use it for baking but try to test your divergent thinking ability and come up with other possibilities. What about as a weapon – you could whack someone over the head with it? Or deploy it as a double-roll toilet paper dispenser, or as fuel on a fire perhaps?

The results were clear: there were significant differences in the levels of creativity among different demographic groups. 'Time and time again adolescents suggested more original and unique uses and came up with more original drawings or geometric configurations than adults. Over time, [the teenagers'] ideas became more original, whether they practised the specific creativity task they had been given or an unrelated task, whereas adults' creativity did not improve,' Stevenson reported. In a related study by Crone, 15–16-year-olds came up with more original drawings of geometric configurations than adults. These studies backed up the work of Sarah-Jayne Blakemore and others into the teenage brain's creativity, and also corroborated studies that suggest that for all of us, creativity, like memory or expertise, can be honed.

There's a fascinating caveat to this particular study, and it's one that points to something crucial for a group's lateral thinking. Stevenson and Crone's team assessed the ideas from adolescents as more original than adults' but noted that it was the adults' ideas that tended to be more feasible. As Claire Stevenson said, 'In the real world, we need people whose ideas are in the sweet spot of creativity that combines originality with usefulness.' We also need diverse teams so that strengths balance out and originality can be aligned with wisdom to produce novel, practical and feasible solutions. 'Perhaps the best approach,' as Claire put it, 'is to encourage adolescents and adults to take advantage of their respective strengths by thinking together creatively about how to solve local and global challenges.' This is how creativity can be embedded in a group and incorporated into their learned behaviour.

Beyond the experts and creatives, every group needs team players, networkers and facilitators in order to be a team of all the talents. How do these other skills arise in the brain and how might we consciously try to recruit them? Robin Dunbar's work on the social brain has an interesting application here. The sorts of skills that support a great networker correlate with the tendency towards being an extrovert. A capacity to work well in a smaller group correlates with a more introvert tendency. Dunbar has identified both of these social styles as aspects of social intelligence with a high genetic component. Those people who find it pleasurable and rewarding to have a lot of social contacts and be involved in many different groups tend to have larger regions within the prefrontal cortex – the area of the brain behind the forehead. The size of these brain regions can actually be used to predict, with a high level of accuracy, the size of a person's social network.

Both approaches to socialising and to group work are valuable and each group finds their own approach rewarding. An introverted person might have fewer friends but be closer to them, for example. An extrovert might be happy to spread themselves a little more thinly. The value in terms of collective intelligence is equal. Those of us who feel more comfortable in smaller, trusted groups will do our best work in these environments and might well perform brilliantly as team players. But small groups get stagnant without a flow of incoming ideas. If there were no extroverts moving between groups and facilitating innovation and cooperation between them, then the group's intelligence would eventually wither. This insight leads to an important principle for anyone hoping to encourage collective intelligence: let individuals and small teams work independently before linking them to one another.

Draw on All the Talents: Neurodiversity

What about the skills and insights offered by people who are not neurotypical, such as those who have been diagnosed with autism or ADHD? In the past these conditions have been equated with disability, and there's no doubt that both conditions can, at their most extreme, cause people significant, even life-limiting problems. But society is starting to appreciate that for some people with less severe diagnoses, these conditions might in fact confer strengths. There is a growing body of evidence to link ADHD and entrepreneurship, for example.

A recent study by Dr Juliette Harris's Twin Research and Genetic Epidemiology laboratory at King's College, London found a genetic link between a variation in the dopamine receptor gene, a diagnosis of ADHD and the tendency to be an entrepreneur. Speaking simply, dopamine is the pleasure and motivation chemical in our brain, tapping into the reward circuit that spurs on many of our more complex behaviours. The variation in the dopamine receptor gene is by no means the only factor involved in the development of ADHD, which is believed to be approximately 80 per cent heritable – many different genes will play their own small part – but it does seem to be highly significant.

The study looked at genetic material from over 1,000 people and discovered something interesting on the DRD3 gene, which codes for a receptor for dopamine that activates the next nerve cell in the pleasure circuit. This variant tallies with a diagnosis of ADHD and has been previously identified as the so-called 'explorer gene' since it confers a tendency to restless and novelty-seeking behaviour. In our evolutionary past this variant conferred an advantage since it drove people on to seek out new solutions to problems. Nowadays, these traits mean that people with this variant tend to thrive in times of crisis and relish the insecurity and uncertainty that come along with exploring a new idea or starting their own business.

There was a 2017 study by Professor Holger Patzelt of the Entrepreneurship Research Institute at the Technical University of Munich (TUM), Germany, into decision-making in people with ADHD that also linked them with innovation and leadership. Although they often find it difficult to suppress impulsiveness, this tends to lead to more intuitive decision-making, which may in fact be a more intelligent way to evaluate decision-making in certain situations. 'The way we evaluate entrepreneurial decisions is largely based on rationality and good outcomes,' Patzelt commented. 'In view of the multitude of uncertainties, however, can such decisions always be rational? People with ADHD show us a different logic that is perhaps better suited to entrepreneurship.'

ADHD is to creativity and entrepreneurship as autism is to expertise, meticulous attention to detail and commitment to a task: behaviours that – unsurprisingly – many employers value highly. Many software companies, including Microsoft, SAP, HP and Relic, have 'autism hiring programmes', in recognition of the fact that some people with this diagnosis have exceptional capacity for logic, analysis and focus.

Autism doesn't only equip people for computer programming jobs, of course. Climate activist Greta Thunberg, who has been diagnosed with Autism Spectrum Disorder, rejects the intimation that it's an illness and calls it her 'superpower'. She views man-made climate change in black and white terms: it requires immediate action and so she acts. This clarity of purpose and singular vision have supported her campaigning work and seen her receive two nominations for the Nobel Peace Prize before her seventeenth birthday.

Greta has a fairly mild version of autism. Her experience is a long way from that of those people I worked with when I was a nursing assistant at a psychiatric hospital, who would not have been able to do what she has done or to integrate into a team. But for every person whose life experience is severely curtailed by their condition, there are many more whose perspective could be valuable and whose working style could make them exceptional members of a group, so long as their contributions were welcomed and their needs met.

Fortunately, attitudes to autism, ADHD and a host of other conditions are changing, thanks to tireless work by campaigners.

There will be more reappraisal of the value of cognitive diversity as the science extends our understanding of these conditions and improves the range of treatments available for them where appropriate.

This reappraisal fits with the biology. If we view conditions of the mind as interesting divergences rather than illnesses or failings, it starts to make sense that they should have persisted over time, through populations across the world. Schizophrenia, for example, can be devastating for those diagnosed and their family and friends, as I know from having studied it in depth and spoken to many people who live with it. And yet even though it reduces an individual's likelihood of having children, its prevalence stays stable all over the world at about 1 per cent of the population.

Studies have shown that there is a strong genetic component to developing the condition, with up to a hundred genes involved. Could it be that at least some of those genes confer an advantage? Relatives of those with schizophrenia tend to be creative. Perhaps some of those genes are linked to desirable intelligence traits like innovation and creative flair, which is why they persist, but when enough of them combine in one person, there is a risk that their impact becomes debilitating.

People who have been diagnosed with schizophrenia often struggle to distinguish between self and other, which means their social skills and capacity to contribute to groups is limited. They also have difficulty with differentiating between reality and illusion. Hallucinations are one of the primary markers of the condition and can be very distressing. But not everybody who hallucinates has a mental health problem. Up to 17 per cent of the population has experienced symptoms of psychosis such as visual hallucinations and many of these experiences are associated with greater creativity and less fixed self-consciousness. We are all living in the world our brains have conjured up for us to inhabit and most of us are delusional occasionally, whether our brains are under pressure from illness, stress, drugs or simply being in love. While persistent psychosis is obviously deeply disabling, perhaps we allow our fear of these conditions to cloud our judgement about the spectrum of experiences. I wonder whether one day we will actively recruit people who have experienced hallucinations, valuing them for their creativity and innovation?

How varied does a group have to be to function well? And can there be such a thing as too much diversity? There does seem to be an optimum range – neither too homogenous nor too eclectic. Just as there is a limit on group size beyond which interactions become chaotic, there is also a limit on the range of perspectives, which must not be too wildly different. Our brains' tendency to dislike the hard work of understanding somebody who is very different from us is always lurking in the wings to sabotage a team.

One possible way to mitigate this is to build in a similarity of competence levels across the group. Our analysis on *The Family Brain Games* showed that a group of four in which everyone is roughly equally skilled (in this example, skilled meant having similar IQ scores) will outperform a group in which there is considerable disparity between the most competent and the least, even if the most competent members are exceptionally skilled.

This finding has been demonstrated in the lab by Professor Chris Frith and colleagues at University College London, who have performed some groundbreaking studies in this area. They found that equal levels of competence, and the ability to freely communicate, helped to amplify the brain power available. But if these two factors were missing, individual brains fared better by themselves.

In one experiment, volunteers came into the laboratory in pairs and were asked to detect a very weak signal that was shown on a computer screen, noting what they saw individually before being able to freely discuss their impression with their partner to arrive at a joint consensus. The results showed that joint decision-making increased sensitivity by around 30 per cent, even when compared with the decision made by the better-performing individual on their own. As Chris explained to me, 'By working together, our subjects could deal with all sorts of problems [more effectively], from dot-counting to logic.' But this drastic improvement of group thought occurred *only* when both partners were more or less equally competent at the task, and could *freely discuss* their disagreements.

The final experiment in the study looked at what happened if one individual in the pair had the wrong information. Chris randomly tweaked the incoming signal to make it noisier and inaccurate for one person in each pair. This significantly impaired their joint decision-making. Chris warned, 'When one person is working with flawed information – or perhaps is less able at their job – this can have a very negative effect on the outcome.' Joined-up thinking doesn't work when a member of the team is incompetent but doesn't know it. We don't always know what we don't know, of course, but the insight from this experiment might give us pause when we're about to dive into a group discussion. We sometimes feel obliged to contribute, but it's worth asking whether we're in a good position to do so or whether our limited knowledge might actually impede the group's success.

In this chapter we've looked at how to put together a team that's well placed to collaborate. We've stressed the importance of recruiting diversity in order to avoid knowledge clustering and clone-like teams. This is worth doing even if it feels counterintuitive. It can be tempting to recruit from the top university school for a specific subject area, for example, but that risks ending up with a team of people who've all absorbed the same ideas and used the same models, potentially amplifying and echoing any biases in thinking rather than bringing in new, creative ways of doing. Better to cast the net further afield. And keep the operation unit at a manageable size. Small groups within a swarm of 150 people help to foster friendships and knowledge exchange, with people getting to know each other's strengths and weaknesses, saving vital time by knowing immediately who to go to and for what.

But even with a diverse team, it can all go horribly wrong! Sensitive and collaborative leadership that avoids expertise bias or egotism is vital, as are active listening and honest communication across the whole group. Otherwise, dominance dynamics can insidiously creep in to amplify biases and steer group consensus in completely the wrong direction. How can we lead, interact and communicate in ways that boost social perceptiveness and so up our teams' collective intelligence? These questions are the subject of our next chapter.

EXERCISE: Sit With Silence Together

At the beginning of a meeting or gathering, highlight any issues that need resolving, then set out the purpose of the group sitting in silence together. Everyone will remain silent for three minutes, which might feel uncomfortable at first but it's an opportunity to embrace that silence. Invite people to let their minds wander, then consider the aim of the meeting and allow any thoughts and associations to come to their minds.

This practice can foster the skill of knowing when, and when *not*, to speak. It also provides a quiet space for contemplation and time for solo brain work before wider discussion.

CHAPTER FIVE

Harnessing Brain Power: Strategies for Smart Teams

We are probably all familiar with the pejorative term 'design by committee'. A camel is proverbially described as 'a horse designed by committee' since its distinguishing features – that humpy back, its foul temper and tendency to spit – feel like a botched compromise made by a team of people with no experience of riding animals across deserts, and no idea how to agree on anything. The expression may be more unfair to camels than committees. Anyone who has sat through enough meetings – whether of the company board, the school Parent Teacher Association or the golf club – would probably agree that they frequently lead to flabby compromise or tense deadlock rather than an outpouring of collective intelligence.

Every time a bunch of people get together to discuss, decide or design something, there is always potential for confusion or conflict. Group work can throw up uncomfortable dominance dynamics where terrible ideas are followed through simply because they originate from the person of highest power. Sometimes, in an effort to please everyone, ideas get averaged out and the result is nonsensical or totally lacking in vision.

So how can we transform the typically messy range of diverse ideas into a meaningful (or even elegant) conclusion? Human beings have invented all sorts of mechanisms for pooling collective intelligence, from the open debate in parliaments to the jury system for trying court cases. They don't always lead to a successful outcome but they are cornerstones of our political systems and our culture for good reason. Human beings are not doomed to conflict or muddle, thankfully.

That said, there is always complexity when a group tries to work together. Collective intelligence won't happen simply through installing your diverse team in an open-plan office with lots of breakout areas. Without systems to support listening, communication, tolerance for ambiguity and conflict resolution, even a diverse team with good intentions is likely to revert to the usual power structures, get mired in disagreements or end up with a proverbial camel on their hands. Their collective intelligence will remain limited.

This chapter looks beyond diversity to inclusion, meaning that we don't merely invite more and different people to a conversation, we also enable them to participate. This is in our rational self-interest because, as we know, diverse groups are proven to be more effective and productive. It also speaks to our collective sense of justice – which is not a mere nice-to-have but a fundamental underpinning of collective intelligence.

There are no quick wins or life hacks in the strategies here. Building collective intelligence takes patience and tolerance of uncertainty and discomfort; it's a medium- to long-term game. Fortunately it also generates some quick wins along the way, as we will see.

First of all, let's look at the crucial role of leadership in creating a culture that supports collective intelligence. The behaviours most needed in well-functioning groups won't spontaneously occur without facilitation. That needs to come from somebody, or some people, with the authority to lay out the rationale for changing the way the group works and making it happen. There's an interesting paradox around leaders and maximising group intelligence. Their role is crucial but it needs to be flexible. Their style needs to be collaborative but their responsibility for decision-making needs to be clear. So what are the particular skills that support a person to carry out this delicate balancing act?

We'll also examine how to avoid dominance dynamics to ensure that every participant is able to contribute, and how to support our groups to be more resilient, since a resilient brain is a smarter brain. Given that trust and openness have been shown to be harder in more diverse groups, which can be uncomfortable and frustrating, we need ways to understand and cope with those feelings. There are many brilliant cognitive strategies for dealing with these difficulties in groups, from reverse mentoring to unconscious bias training, group mindfulness sessions to using brain*writing* rather than brainstorming. Crucial to everything though is open communication and clear leadership. Let's start there.

Leading for Collective Intelligence

How do we organise our teams for maximum collective intelligence? Do we need leaders at all? Can we not just assemble a smart group and empower them to self-organise?

Over the past two decades many organisations of all stripes have opted for flatter structures, partly in response to the communications technologies that have made it easier to share information and ideas, partly because behavioural scientists have been highlighting the benefits to group work for many years. Flatter groups encourage more autonomy at lower levels. People in positions of responsibility tend to have more colleagues reporting to them, so groups are multidisciplinary. This is definitely more conducive to collective intelligence than the old-fashioned model of a strict hierarchy with power concentrated right at the top. When only senior managers were expected to come up with good ideas or take strategic decisions, collective intelligence was so limited as to be non-existent.

But a flat organisation, like a diverse team, is no guarantee of anything. It can only take you so far. It's certainly not as simple as saying 'the flatter the better'. Consider Google. In July 2001, with millions of users and over 400 employees on the books, its founders, Larry Page and Sergey Brin, decided to get rid of the entire management team. Page saw the layers of leadership as stifling to Google's innovative engineers. He argued that since they hired only the most talented people, supervision was a hindrance rather than a help. Some engineers thrived under this new approach but generally the company was thrown into chaos and confusion. Without managerial support, problems didn't get solved, projects didn't get resources and engineers went without feedback. Within a year Google had reverted to the old system.

Now, this story might sound like an illustration of the blindingly obvious – groups need *some* level of accountability and supervision otherwise chaos takes over – but that is partly the power of hindsight. After all, if nobody ever tests out the crazy ideas, stagnant thinking will eventually doom us to collective mediocrity. Besides, what level of direction is optimum? A flexible leader within a flat-ish structure, perhaps?

The more we aspire to collective intelligence, the more – paradoxically – we need some sort of leadership at key moments during projects. We need leaders or influencers to define the value of collective intelligence to the group at the outset. We need them to support the pro-social values the group depends on. We also need them to be highly flexible: capable of letting go of power but then shifting back into responsibility when necessary. The first requirement of intelligent leading for collective leadership is to co-create a culture of shared objectives, shared information and shared success. Without that, conflict is all but inevitable. This turns out to be true not just for humans but even for robots.

In 2017, computational social scientists at the Internet Institute, University of Oxford and the Alan Turing Institute, London published a study on the behaviour of editorial bots that maintain the pages of Wikipedia, the online encyclopaedia. They analysed a decade's worth of data, (from 2001 to 2010) from thirteen different language editions, and uncovered startling patterns of conflict among rival bots. The bots had been programmed to be autonomous and to carry out various tasks such as undoing vandalism, enforcing bans, checking spelling, creating links and importing new content automatically. The scientists assumed, before they began the study, that interactions between bots would be 'relatively predictable and uneventful'. What they found, however, was a battleground. Rival bots were engaged in long-running editorial fights, undoing each other's edits over and over again and getting completely caught up in this conflict rather than doing their work.

In 2021, I asked Taha Yasseri, one of the study's authors, whether the fights were still raging. He told me that things had calmed down considerably thanks to years' worth of tweaks to the bots' programming to make them more interdependent and less autonomous. The issue, as Yasseri described it, seemed to be that nobody had anticipated that the bots' lack of social intelligence would be a problem. They'd been set up as free agents empowered to do their own tasks; and individually, each was highly competent. The programmers hadn't imagined that without direction, the bots would struggle to assess how their tasks fitted in with the tasks of others.

The Wikipedia bots were in thrall to their own individual agenda and incapable of taking on board that other bots might have different but equally valid agendas. They could not have intelligent interactions rooted in pro-social behaviour, so clashes were, with hindsight, inevitable. A robot team leader capable of directing, mediating and acting as referee would no doubt have been a great help but AI is still far away from being able to carry out these non-analytical cognitive functions. Human beings have a vast advantage here, which is fortunate because collective intelligence can only flourish in a group where the leader or leaders are facilitating it, and where all members of the group have, in effect, signed up to support it.

Leaders may be necessary to make the case for collective intelligence and to implement the strategies for it to emerge, but they also need to be able to step out of the limelight so it can actually happen. An individual or a clique that wants to hang on to power for themselves is a huge block on collective intelligence. We need a particular kind of leader, or leaders, to facilitate without dominating. We need leaders with fantastic social and emotional intelligence.

Power and Consensus

As soon as we start to talk about 'leaders', many of us begin to make assumptions. The word evokes a certain kind of group – a corporation or a political party perhaps – and a certain kind of person. One infamous headline a few years ago reported that one in five CEOs were psychopaths. But is leadership always synonymous with individual power? Does the bad guy always claw his way to the top? The latest neuroscience suggests not, and that's good news for those of us who believe there's more potential in collective intelligence and flexible leadership than in any hierarchy, even if the CEO is a genius.

Broadening our understanding of leaders and leadership challenges us to think in a more collective and less individualistic way. It pushes us to go beyond our own biases and think afresh. All of us who have any kind of influence over other people are in our own way leaders. Parents lead families. A particular child (never our own) leads other children astray. There are community leaders, who may or may not have been elected. Leadership can be informal. It need not last long. At my son's recent birthday party, a fellow parent stepped up to mediate between a group of squabbling children. She was the leader that group needed at that moment.

What do successful leaders do to foster greater joined-up thinking in the groups where they exercise influence? And what skills do they themselves need in order to do it? Looking at the way leadership occurs in groups of other animals and in groups of humans from different cultures can help to answer these questions. It's not quite as simple as saying that authoritarian leadership styles are always bad. Just as groups benefit from being organised around the sweet spot of a flat-ish structure, so they also benefit from leadership that is flexible and agile enough to shift in and out of different modes as the situation demands.

That parent who stepped up, supported the group and then stepped down again was doing something that we see in groups of animals all the time. It's called 'transitory leadership' and it's an immediate and temporary response to a specific problem. Animal leaders are often transitory. A pack of hunting wolves, for example, will typically have one mating couple in their leadership team who share the jobs of navigating, leading the hunting and child-rearing. The larger the pack, the more fluid this hierarchical structure. Different couples step in and out of this role without conflict, as the situation demands and depending on their capacities.

Leadership shifts among the members of a murmuration of starlings in a similar way; it is not fixed with one member for the entire migration. It's just too exhausting for any one individual to lead for the whole journey, so it makes sense to have a tag team of refreshed leaders ready to serve in the wings. The transitions are driven by the needs of the group and the knowledge and the physical capacities of individuals – not by leadership battles.

Adjunct Professor Jens Krause, an expert on swarm intelligence who heads a laboratory at Technical University Berlin, has conducted experiments using robotic fish to study leadership and group decision-making. 'In what we call open groups [with flexible membership – this includes many fish schools and bird flocks], leadership is a question of information status, not rank or dominance. One fish or bird may notice something in the environment that is relevant and its behaviour will influence that of others in the group. When another individual discovers something, the influence of the first one wanes and the "new" leader takes over. This form of self-organised, information-based leadership is very common and allows for a smooth transition of leadership.'

Human beings also have transitory leaders in social groups, politics and business, of course, but we are sometimes less graceful in our ceding of power than the wolves or the starlings. Jens's research corroborates that humans are capable of adopting this sensible strategy and are generally more successful when they do. It's egos that sometimes get in the way. An individualistic culture that strongly associates leadership with personal power and prestige sets an individual's needs in conflict with the needs of a group, even though the data doesn't bear this out. So the extent to which a leader has the social intelligence to think 'group' rather than 'self' correlates directly with their personal capacity to drive and foster joined-up thinking.

The second crucial skill is the ability to think and organise flexibly so that the group can harness as many of the benefits of both hierarchies *and* open groups as possible. Roles and objectives must be clearly defined, for example. Open groups with transitory, expertise-led leadership tend to increase the responsiveness of the collective, build loyalty and harness creativity. The pitfalls of moving leadership too far either towards hierarchy or openness are worth bearing in mind. In some cases, they can be a matter of life or death, as a study into Everest climbers shows.

In 2015, Professor Adam Galinsky and colleagues from Columbia University, New York and INSEAD, the non-profit business school based in Switzerland, ran a fascinating and slightly terrifying study into which groups of mountaineers were most likely to make it to the top of Everest, and how that tallied with individual fatality rates. They analysed data from more than 5,000 expeditions, looking at the impact of nationality and how it related to cultural values around hierarchy. They tallied nationality with two different global inventories of these values and tried to control for factors such as weather, age and experience of climbers. Their results were stark. Teams from Russia, China and other countries where the culture is more hierarchical and the leadership style is characterised as authoritarian were much more likely to make the summit than teams led by individuals from countries with more collaborative cultures such as France or the USA.

Unfortunately, they were also more likely to die. In a situation that exposes people to severe risk, communication was fundamental and so was decision-making. There was no doubt that having a leader was essential, but so was empowering the lower-ranking members of the team to be able to speak up if they believed they had information on changing weather conditions, say, that had been overlooked and needed to be fed into the decision-making process. The study concluded that there are big benefits and serious costs to hierarchies, and that group leaders should be aware of them and aim consciously for the sweet spot between them.

One way to do this is by assembling a team that blends different areas of expertise *and* different levels of experience and specialisation. The team has one single overall leader but tasks are broken down into segments and leadership on these segments is distributed around the group. This is an approach used widely in surgery teams, where the lead surgeon has overall decision-making control but the head nurse is in charge of the checklist of safety procedures before and after surgery.

The myth of the 'psycho' CEO and the Pro-social Alternative

So far we've identified an ability to prioritise the needs and welfare of the group, an understanding of the costs and benefits of different ways to lead *and* the flexibility to shift between them, as key skills for leaders of intelligent groups. What about more personal qualities?

Researchers Cameron Anderson and Oliver John from the University of California, Berkeley have conducted a decades-long study into whether a hierarchical corporate work environment and culture propels people with marked antisocial personality traits to the top. The researchers took 671 American undergraduates from three different universities and analysed them with personality assessments, rating them for disagreeable attributes. Fourteen years later, when their professional careers had unfolded, they assessed them again to figure out the degree of power they had attained at work. Not wanting to rely entirely on self-reports (there are dangerously deluded self-important people out there, as well as self-effacing humble ones), the researchers spoke to the study participants' co-workers to fact check their accounts and find out how they were to work with.

The findings, published in 2020, were clear: being manipulative or abrasive *did not* lead to greater success. Those who were generous, trustworthy and generally nice were just as likely to have obtained power as those who scored high on disagreeable traits. 'I was surprised by the consistency of the findings. No matter the individual or the context, disagreeableness did not give people an advantage in the competition for power – even in more cut-throat, "dog-eat-dog" organisational cultures,' said study co-author Cameron Anderson. He and his team point out that although disagreeable individuals can be intimidating, which can elevate power position, their poorer interpersonal relationships offset any advantage that provides.

The key predictive attribute for success turned out to be . . . extroversion. People who are assertive and have high energy tend to do well in corporate structures, which the authors attributed to their sociability. As we've seen, both introverts and extroverts are crucial to intelligent groups but it makes sense that the extroverts' drive and ability to listen to others, take onboard their perspective and act as a bridge between groups might position them as good leaders, especially in today's corporations.

After all, thinking about leadership and leaders has changed a lot over the past few years. Just as hierarchies have been flattened, so we've absorbed the business case and the social need for a more diverse range of leaders. The buzz over the last few years has been around 'transformational leadership', where leaders encourage, inspire and motivate others in the team so that they can contribute effectively, innovate and create change. This style of leadership depends on being able to create an environment of emotional and psychological safety that allows people to speak up and participate fully, and is strongly associated not only with extroverts but also with women.

We've already looked at the link between women and collective intelligence via Professor Anita Woolley's work, which established that the strongest predictor of a group's collective intelligence quotient is the number of women in the group. The suggestion is that women's typically high fluency in pro-social behaviours rooted in empathy explains this. Although as Professor Woolley told me, scientists are still trying to unpick the precise mechanisms of emotional intelligence. 'A number of studies show a negative correlation between testosterone and empathy (or social perceptiveness) . . . which suggests that it is partly biologically based, although of course socialization always plays a role as well.'

Emotional intelligence might also underpin women's capacities in transformational leadership. As Professor Jean Decety from the Psychology Department, University of Chicago puts it,

'Empathy has evolved in the context of parental care for offspring, as well as within kinship bonds, to help facilitate group living.' In other words, given that women have been primary caregivers throughout human history, they may have acquired expertise in skills that rely on picking up on the emotions and needs of others. They may be better than men, *on average*, at creating a context that feels safe enough for people to speak their minds and offer their opinions.

But it's not as if these skills are written out of the brain by the Y chromosome or that men are incapable of empathy or emotional intelligence. Human beings are, generally speaking, empathetic, social animals and people of all identities have brains that are far more alike than they are different. Learned behaviour is always crucial.

It seems clear that these pro-social skills, wherever they come from, underpin meaningful participation in many different areas of life. We need to nurture, exercise and reward the innate skills we share, to help support the next generation. Especially in light of their biology. It does boys a disservice when we don't encourage them to develop their talents in empathy, intuition, communication and power-sharing. They end up with fewer of these crucial cognitive skills and that has an impact – not least on their ability to be a transformational leader!

Empathy is not the only cognitive skill that underpins intelligent leadership of intelligent groups, of course. To return to Anderson and John's recent study, it was extroversion that was the strongest predictor of corporate success. Being an extrovert is a trait, not a skill, but it is strongly associated with certain behaviours that impact on leadership. Extroverts facilitate the sharing of ideas and skills that collective intelligence relies on. They're human pollinators, the equivalent of bees tracking pollen from flower to flower. So is the ideal leader for groups who want to maximise their collective intelligence an empathetic and extrovert woman? Maybe!

This brings us right back to the question of diversity that we already know is crucial for collective intelligence. The problem is, there are still too few women in positions of leadership. In June 2019, the *Harvard Business Review* highlighted the fact that women account for fewer than 7 per cent of the world's national leaders and 5 per cent of Fortune 500 chief executive officers, with these numbers declining globally. Whatever the complex reasons for this, it is clear that we are not harnessing our collective capabilities.

How to tackle this? Studies such as the one carried out by Brian Uzzi, professor of leadership at Northwestern University, have shown that we're stuck in a catch-22 situation. It's hard to achieve diversity of leadership (or participation) without the role models available for mentoring. Women, for example, benefit disproportionately from having mentors who are themselves women. Statistically speaking it's not enough for women to be bright, well qualified and well connected and have a predominantly male inner circle of champions and mentors. All that would still give them a lower job position than having all those advantages and a mostly female group of supporters.

This is at least as true for other social groups who are not traditionally visible in positions of influence. We need high-ranking women, people of colour and people with disabilities in pivotal leadership positions in order to help support the next generation of leaders, and we need them at the table as soon as possible, so that we have a broader range of cognitive power to help problem-solve for the increasing list of global challenges.

Leaders are important; they enable our society to progress and prosper. They can be innovators, agents of change and icons of success. Take the relatively innocuous challenge of running the four-minute mile. This goal had been chased by competitors across the globe from at least 1886 until, on 6 May 1954, Roger Bannister finally managed it. He was a full-time student who had devised his own training regimen and he performed this feat on a cold day, running on a miserable wet track at a small and insignificant race in Oxford.

His extraordinary achievement seemed to open a window of possibility. The following month, John Landy, an Australian runner, ran the four-minute mile in 3 minutes 58 seconds. Then, just a year later, three runners broke the four-minute barrier *in a single race*. Since then more than a thousand runners from around the world have replicated and bettered Bannister's achievement.

Great leadership changes the collective consciousness. When aligned with a pro-social mindset focused on collective success, it could spearhead the technological and behavioural revolutions that will help our species to survive the next age of challenges. Or more simply, it could propel somebody to step up and help a group of children learn the basic skills for collective intelligence at a birthday party. It's all useful, and all needed.

If we broaden our focus and assume we've assembled a diverse group of people, figured out that we need a flat-ish structure and a leader or leaders selected on the basis of their strong emotional and social intelligence, what's next? What can cognitive science tell us about the processes that support collaboration? How can we make sure we're capturing the diverse intelligence available to the group? There are going to be disagreements and impasses because an intelligent group requires diversity of opinion, and that diversity (sometimes) brings conflict. What can be done to maximise the benefits of debate while minimising the downsides of disagreement? How can we dial down our natural discomfort with the uncertainty of interacting with people who are different from us? Everything starts with open communication, but for that to happen, individuals need to feel safe.

Group Dynamics, Dominance and Turn-taking

Collective intelligence is absolutely dependent on the flow of talk, ideas and feedback between members of the group. There are many variables that impact this, from having a shared language to trusting one another enough to open up. Creating psychological safety is crucial to everything, as we know, but what other processes and practices make it easier to work together towards our goals?

Clarity of purpose and commitment to the principle of group work are essential. The leader of the group has a responsibility to emphasise both things. They also have primary responsibility to facilitate conversational styles for the whole group that encourage participation and avoid dominance dynamics. When dominant members in a group express views based on partial information, it can seriously skew the group's thinking. Other participants start to withhold potentially valuable ideas (consciously or unconsciously), for fear of looking stupid or falling out of favour either with the boss or with the group. Knowledge is squandered and the group's collective cognitive bandwidth shrinks. The problem is worse if the dominant figure or figures are also senior members of the team but, even if they're not, the effect on creativity, decision-making and problem-solving can be chilling – and all the more difficult to deal with because the perpetrators are often unaware of what's going on.

In his book *Rebel Ideas*, author Matthew Syed quotes Leigh Thompson, professor of dispute resolution and organisations at the Kellogg School of Management, Northwestern University. She put it in bold statistical terms. 'The evidence suggests that in a typical four-person group, two people do 62 per cent of the talking and in a six-person group, three people do 70 per cent of the talking. It gets progressively worse as the group size gets bigger . . . Perhaps the most remarkable thing is that people don't even realise they are doing it. They are adamant that everyone is speaking equally and that meetings are egalitarian . . . If you point it out to them, they bridle and you get into an escalating conflict.'

Dominance dynamics happen in every setting and in groups of all sizes. They are not merely the preserve of boardrooms or people who have consciously set out to control a group. It's not easy to tackle them, precisely because they surface unconsciously and can be hard for the dominant group members to acknowledge and correct. Leaders in particular need to cultivate self-awareness and sensitivity. It helps for everyone in a group to know about the phenomenon and to acknowledge how important – and difficult – it is to eradicate. This in turn depends on individuals' emotional intelligence and high levels of trust. Not for the first time, we see that all the behavioural components of collective intelligence are mutually dependent.

There are practical strategies that help. Formal conversations such as those that take place in meetings can be carefully facilitated (on a rotating basis – not by the same person every week) to ensure that no single individual speaks beyond a certain time limit and that people wishing to make a point are not invited to speak in decreasing order of seniority.

Mixing up the range of voices we hear from in meetings sounds obvious but can be hard to implement. Cognitive fatigue is a very real phenomenon and there is often a desire to arrive at decisions quickly in the name of efficiency; these issues combine to induce a lack of patience for 'too much chat'. While nobody wants to sit in interminable meetings, good timekeeping needs to be balanced by the need to reap the benefits of cognitive diversity. Asking junior or more introverted staff to speak first, when the meeting still has the collective bandwidth to hear them, is a simple way to combat the desire to grab at solutions before we've heard the full range of views.

This tactic is one of many used at Amazon that has been

credited by founder Jeff Bezos as being fundamental to the company's success. All meetings over the past decade have used the same strategy: the most senior person in the room always speaks last in order to avoid stifling debate.

Deliberate disruption of a group's hierarchies of seniority is key to many strategies being used in business of all sizes, as well as community groups. We see it in the policy of reverse mentoring, where a junior member of staff mentors a more senior person. This is one way to ensure that those desirable but sometimes difficult-to-organise conversations between different people actually happen. If a senior member of staff approaches a junior colleague with a request for help in a particular area, the benefits can be huge and mutual. The senior person gets to hear about the latest trends or tune up their skills in a new area of social media, say. They can briefly be a beginner again and learn something new, with all the benefits for neuroplasticity and neurogenesis that learning implies. The junior person is exposed to their expertise in a context of mutual respect and openness. Trust and loyalty deepen.

Steering Away From Groupthink

Aside from flipping hierarchies in order to avoid dominance dynamics, there are other practical ways to make sure we don't miss out on innovative ideas or arrive at premature consensus. One of the most repeated findings in studies of successful groups is that they allow for periods of individual work as well as communal effort. The most effective joined-up thinking makes space and time for individual expertise and only *then* finds a way to synthesise the results. Collective intelligence relies on individuals who are empowered to think independently without being primed by others, and only then come together to pool ideas and information.

Amazon makes use of this principle, too. There is a period of 'golden silence' at the beginning of every meeting, when everybody reads the agenda and reflects on it. Responses are then gathered verbally (with the most junior person going first, of course) or through brain*writing* rather than brainstorming.

There are numerous variations on this technique but the essential element is that people are given a certain amount of time to write down their ideas on Post-it notes or cards, rather than call them out. There are two advantages to this. The first is that it cuts down on dominance dynamics, where the most senior or confident person sets the tone for everyone else. The second is that a quiet period of ten or fifteen minutes for individual reflection means ideas have time to come up and develop before entering the fray of discussion.

At the end of the writing period the cards can be handed to a facilitator who reads out the ideas in turn and manages the subsequent group discussion. Since cards are anonymous and the facilitator has not taken part in the exercise, this gives a high level of protection for each idea to be evaluated on its own merits. In the more interactive version, cards are passed one at a time to the next participating person in a circle, who can either modify the idea, use it to trigger another one or pass it on to the next person in the group.

This interactive version of brainwriting is particularly powerful at generating both quantity and quality of ideas and solutions but since it's not anonymous, it depends on a well-established context of emotional safety and bonds of trust between group members. Either option is worth exploring. A study published in 2015 by Marcela Litcanu from the University of Timişoara showed that brainwriting generated twice the volume of ideas as well as higher-quality ideas, when rated by independent assessors.

Resilient Brains Are Much More Intelligent

Stress, uncertainty and conflict all undermine collective intelligence by inducing fatigue and increasing anxiety. Any group that wants to nurture joined-up thinking cannot afford to overlook the impacts that erode their team's resilience. Taking care of people is no frippery. A resilient brain is quite simply a smarter and more productive brain. Long-term stress kills off cells in the hippocampus, contributing to the deterioration of memory. Chronic pressure can decrease neuroplasticity and put our so-called 'hot and cold cognition' out of whack. This causes the brain's fear centre, the amygdala, to become hyperactive and hypersensitive at the expense of the insula and cingulate-cortex areas, which are involved in longer-term thinking and horizon scanning. So we get stuck in fire-fighting mode rather than working on new ideas or ways to collaborate on problems.

A little bit of stress can in fact be good for motivation and mental agility, as the manager of the escape room experience pointed out to us when we undertook the challenge. The stress hormones cortisol and adrenaline fire us up to think flexibly and focus harder in order to meet the threat that has raised their levels. Any benefits are short-lived, though, and if stress is frequently repeated or lasts a long time, the effect on all our cognitive capacities, from memory to empathy, can be devastating.

Groups need time in between projects to rest and recuperate. Big tech companies like Google provide exercise and social activities on site because of the proven benefits for individuals, the group, staff loyalty and the bottom line. Physical exercise is the only way we know of to drive the growth of brand new brain cells in the adult brain, as well as being proven to combat depression and fatigue. Social interactions reduce loneliness at work as well as boosting chances for innovation.

Extinction Rebellion, a very different organisation, has a similar philosophy for similar reasons: because it's good for the individual and sustains the work of the whole. The movement has been built on a regenerative culture that seeks to avoid activist burnout and prioritises people's wellbeing. Periods of activity are interspersed with fallow time when groups are encouraged to exchange feedback, learn from one another and rest.

Stress and burnout are not the only ways our resilience gets eroded and our ability to function suffers. Uncertainty and disagreement are other cognitively exhausting states that can sap individuals' energy and undermine collective intelligence. Uncertainty is of course a fact of life but it is particularly prevalent in situations that are dynamic, complex, unpredictable and where we have something to lose or something at risk. Group projects are typically all of these things at once. For as long as the outcome of your team's bid or the launch of a project you've been working on for months is unknowable, you're experiencing uncertainty.

People respond to this in different ways. Some of us might feel it as a low background rumble of worry, or as the jitters of excitement. Others find it much harder to tolerate and become stressed, paralysed by indecision, snappy with colleagues or anxious for ways to resolve the uncertainty even if they're not conducive to a successful outcome.

Greater social and cognitive diversity in a group exacerbates uncertainty, leading to a corresponding increase in the amount of friction between team members. As we know, our brains are inherently lazy and biased and we are hard-wired to find it easier to understand and get on with people we perceive as similar to us. So when a team's diversity goes up, we need to put in place strategies for increasing people's ability to tolerate ambiguity, uncertainty and conflict.

Cognitive scientists have developed a scale of Uncertainty Intolerance, which they use to measure people's differing reactions on a spectrum ranging from feeling buzzed to feeling panicky.

Having a high intolerance for uncertainty means that the person perceives it to be intrinsically threatening. This leads to feelings of being overwhelmed, or wanting to get out of the situation or impose certainty at whatever cost. The amygdala is once again in control at the expense of the anterior-cingulate cortex.

A high intolerance for uncertainty is not a disorder so much as a personality trait, but it has been identified as a contributing factor in a number of mental health problems including anxiety disorder, emotional disturbance, obsessive compulsive disorder (OCD) and eating disorders. High *tolerance* for uncertainty, on the other hand, is associated with increased flexibility of thinking and greater capacity to manage emotions, approaching new or uncertain situations with curiosity rather than alarm and seeking opportunities to learn, or do things differently.

As with all the complex behaviours that make up our personalities, our capacity to tolerate uncertainty is not a fixed characteristic but a bundle of tendencies that we can tweak and train. We could probably all benefit from boosting it but it's especially important for people looking to expand the diversity of their groups and make the most of their collective intelligence. So how do we do it?

Layla Mofrad is a psychological therapist who researches the effects of uncertainty on mental health and is based at Cumbria, Northumberland, Tyne and Wear NHS Foundation Trust. She emphasises that there is nothing unusual about finding uncertainty uncomfortable, especially if it affects the fundamentals of our lives or the people we love, and if it goes on for a long time. Mofrad and her colleagues at the Uncertainty in Contexts Research Network began their study during the coronavirus pandemic when many, perhaps most, people were finding it hard to cope with not knowing what was coming next. But problems can arise when discomfort is so high that it makes us turn to unhealthy ways to cope with it, such as avoiding anything that makes us feel uncertain, or trying to control other people.

'People with higher intolerance for uncertainty tend to fall into two camps when it comes to strategies for dealing with it,' Layla told me. 'They either under-engage with the problem or over-engage with it. A lot of the work we do at our clinic encourages under-engagers to extend their comfort zone by engaging a little more and over-engagers to do a little less. The aim is to build people's capacity to feel the distressing feelings, without reacting. Over time, this increases their resilience.'

Mofrad has developed a playful, low-pressure approach to increasing people's capacity to handle uncertainty. She creates situations where there is a very small amount of uncertainty by inviting people to play games such as Jenga, where the tower will fall at some point but nobody knows when. Some other strategies, such as turning the car's satnav off when travelling a familiar route or ordering something different at the takeaway, are carried out alone but there is something particularly powerful about the work in the therapeutic group. It normalises the feelings for people and allows them to learn from and support each other. And because interaction and debate with others is inherently uncertain, the simple fact of working through exercises together means there is a stretch beyond comfort zones but it's happening in a place of emotional safety.

These are precisely the conditions that we all need to create in order for our groups to function intelligently. Small risks that might lead to big discoveries can then be taken because group participants understand that they are valued and will not be punished if the risk doesn't pay off.

Smart Ways with Conflict

Mofrad's work also speaks to strategies for handling the discomfort that can come from disagreement. Effective group work cannot and should not avoid difference of opinion, as we know.

Amy Gallo, a contributing editor at the *Harvard Business Review*, describes debate as a process of generating 'creative friction [that] is likely to lead to new solutions'. But while debate is welcome and productive when carefully facilitated, it can shade into disagreement and even conflict, which definitely does undermine collective intelligence.

It's not always possible to delineate clearly between debate and disagreement. Some people find the exchange of ideas energising; others shy away from it. Different people can perceive the same conversation in very different ways. One person's constructive criticism is another's outright hostility. This difference in perception will depend on many factors, from the individual's tolerance of uncertainty and the sensitivity of their amygdala to a whole host of environmental and social considerations.

It is also inflected by status. Cameron Anderson, professor in leadership and communication at the University of California, Berkeley, conducted research into different types of disagreement in groups. He found that when two people involved in a disagreement both viewed themselves as higher in status than the other, it was more likely that they would reduce their contributions to the group and focus their energies on winning their private battle. Conflicts of this kind between members of self-perceived higher status can substantially decrease group performance.

Anderson's suggestion was that creating little bubbles of expertise, where individuals are clearly delineated in terms of their skill rather than segregated by hierarchy, would reduce status squabbles and lead to more effective group functioning and enhanced team performance.

Some disagreement is inevitable in healthy and intelligent groups, but it can be especially uncomfortable and frustrating if we suspect that someone is not acting in good faith. We've probably all encountered the sort of person who appears to relish being deliberately provocative or even disagreeable, but it might help to remember that they are a definite minority. In general, people are more likely to find agreement than disagreement rewarding at the neurochemical level. Experiments by Chris Frith and colleagues at University College London show how reward circuits light up in people's brains when they discover that they agree with one another. Most of us, most of the time, are looking for ways to collaborate.

There are ways we can dial down the potential for conflict while welcoming difference and debate. The group's leader can set the tone, reminding members that intelligent groups don't stifle conversations but do steer away from conflict. They can ask everyone to put the group's shared purpose and values of mutual respect into practice. The leader and the group share responsibility for creating the framework of emotional safety for all. When this expectation is embedded in group dynamics, it makes conflict much less likely and less extreme when it does occur.

We can also take personal responsibility for the dynamics we bring to the group. We often overreact to feelings of discomfort and try to clamp down on them immediately. Mofrad suggests we practise opening our minds and bodies up to them through breathing exercises and gentle stretching. This can slow down the fight-or-flight response and support our brains to reassert their executive functioning. Boosting our sense of internal emotional safety ahead of a meeting that we know will be tense, for example, can increase our resilience and our patience. Any activity we find soothing will help, from eating a favourite comfort food to taking a long, hot bath. This conscious self-soothing works well in the aftermath of conflict as well.

Some conflicts, whether personal or political, are just not amenable to harmonious resolution. Later in the book we will look in more detail at what happens when collective intelligence is corrupted by an individual or group's will to power, or by emotional contagion. Some people, such as psychopaths and narcissists, have particular brain profiles that render them incapable of participating in joined-up thinking.

For now, it's worth remembering that if there is bullying or abuse occurring, we are under no obligation to stick around and try to mediate our way out. Occasionally, the best way to resolve conflict is by removing ourselves from it. There is a strand of thinking within evolutionary biology called movement chauvinism, which says that human beings evolved because we moved about – on the hunt for food and other resources and seeking out novel ideas. If a situation or environment wasn't working, our restless novelty-seeking brains took over and moved us on. So the next time you encounter persistent tension or conflict, consider taking a leaf out of the movement chauvinists' book and change your environment by simply walking away. This might mean leaving a situation permanently – can you change your job if the team is toxic? – though if this isn't possible, it might be sufficient to make a temporary change. Leave the room, get some space and come back later.

There are other ways to disengage from conflict if physical movement isn't an option. Withdraw your attention and send it elsewhere so that you're distracted. Lean into your capacity to feel curiosity rather than dislike. Remember that difference can be productive, even if it's uncomfortable. Trust your intuitive intelligence and your embodied cognition to help your brain come up with a good solution.

Boost the Range of Emotional Intelligence

As well as empathetic awareness, there are other cognitive skills relating to emotions that we depend on such as levels of interoceptive awareness, which is a huge and exciting new area of cognitive research that we'll be looking at in the next chapter. The ability to contain and control our own emotions is crucial, as is being able to spot when other people are either out of control emotionally or are deploying their emotions in a strategic way.

High levels of emotional intelligence are, as we know, fundamental to collective intelligence. High susceptibility to emotional contagion, on the other hand, can potentially paralyse a group with anxiety and leave them vulnerable to manipulation. Human beings are prone to synchronising emotions, and this has a huge effect on group behaviour. So while we definitely want to encourage our groups to build their awareness of emotions, part of that involves nurturing an understanding that emotions are not always benign and that, just like ideas, they can be spread easily.

Emotional contagion has become much more visible since social media has meant we can all see its effects playing out in real time, as positive or negative content sweeps through our newsfeeds. But long before Facebook was thought of, cognitive scientists knew that emotions could be shared via language as well as spread through non-linguistic behaviours and hormonal cues.

Back in 1986, behavioural psychologists Peter Carnevale and Alice Isen of the University of Southern California carried out a pivotal study on the way that emotions affect behaviour, specifically negotiation and collaboration in pair groups. There were forty pairs involved in the study, half of whom were manipulated into a positive mood by being offered sweets and funny cartoons to read before the experiment began. The other half did not receive positive treatments and were told to sit in a blank environment and wait briefly. All pairs were then given a negotiation task to complete.

The results were striking: those who had been primed to be in a happier mood exchanged information and cooperated to come up with creative solutions to the task. They were likely to reach an outcome that allowed both sides to settle for something beyond a 50–50 split. Those who were not primed with positivity were far more likely to become competitive, uncooperative and hostile, which resulted in poor bargaining outcomes. These results have been borne out many times since: when members of a group feel positive, they are much more likely to work collaboratively and come up with creative solutions to problems.

Emotional priming doesn't need one-on-one interaction to be effective. In fact, it doesn't even need to be sincere emotion to have a contagious priming effect. Dr Sigal Barsade, professor of management at the Wharton School, University of Pennsylvania, studied the impact of emotion on decision-making in the business world. She ran a study in which a number of people were divided into four groups and told they needed to act as salary committees with responsibility for deciding how to allocate funds. Each group was given the same presentation, inflected with a different emotion. Presentations were delivered by an actor who'd been assigned to convey one of four different moods: enthusiasm, warmth, irritability or sluggishness.

The actor's entirely simulated mood not only spread among the committee members, it also affected the way the group worked together and the decisions they reached. Groups in which the actor conveyed positive emotions cooperated more, argued less, and were ultimately more satisfied with their performance than groups in which the actor had conveyed negativity. Given that actors were able to influence group behaviour without the group's awareness and with fake emotions, this study suggests that deliberate emotional contagion offers considerable potential for unscrupulous manipulation. We'll be looking in much more detail at groups' cognitive vulnerabilities and the potential for collective stupidity in Chapter Seven.

So, given that we know emotions – both positive and negative – make a substantive difference to performance, what can we do to build our awareness of their effects? On the positive side, Dr Barsade advocated that enthusiasm and gratitude should be the cornerstones of everybody's approach to their interactions. Leaders and influencers have a particular responsibility to do what they can to prime their groups positively rather than negatively. Saying thank you to group members for effort as well as success, for example, has been shown to be massively beneficial to morale. In 2010, Professor Adam Grant at the University of Pennsylvania, and Francesca Gino, associate professor at Harvard Business School, found that it helps individuals to feel socially valued and promotes widespread pro-social behaviour since it primes people to think in terms of the group rather than themselves. We can all use the power of emotional contagion consciously. The mantra should be: convey the emotions you want to inspire.

Aside from enthusiasm and gratitude, it can also help to be mindful of how we give feedback. We need to ensure it's presented not as an assessment but as a tool for individuals and groups to learn from their past experiences.

Carol Dweck, professor of psychology at Stanford University, has conducted research that shows how different styles of feedback to a child on their academic achievements can alter their future intelligence scores. Telling the child that they are intelligent and praising them for this 'intrinsic' quality indicates that their intelligence is fixed and not necessarily responsive to effort. This style of feedback frequently has the opposite effect to the one that was desired, since it encourages children who perceive themselves as bright to think that they don't need to work hard, and children who perceive themselves as less bright to believe there's nothing they can do to improve. It's demonstrably better for a child – or anybody – to believe they can develop their skillsets. There is a huge body of research by Professor Dweck and many others to suggest that a growth mindset is a strong predictor of ongoing success. This is true for groups as well as individuals. A positive and encouraging culture based on curiosity, security in experimentation and the value of learning is paramount to boost collective intelligence. But, as we've seen, it's also worth being aware of the power of emotional contagion to undermine a group's purpose, mutual trust and cohesion. This can happen when a charismatic individual or group is allowed to set the emotional mood.

Charisma is another of those personality traits that can be hard to define but is easy to recognise when we encounter it. Some people are just very accomplished at putting others at their ease. They're witty, attentive and charming, great entertainers, compelling conversationalists, creative or innovative in their thinking.

Recent work into the neurological basis of charisma suggests it is made up of elevated emotional intelligence – an ability to assess dominance dynamics and quickly read the mood of a room – combined with mental agility or quick-wittedness. In studies of more than 400 participants published by William von Hippel and colleagues at the University of Queensland in 2015, individuals were rated for how 'charismatic', 'funny' and 'quick-witted' they were (as judged by their close but critical friends, rather than independent assessors). They were then asked a rapid-fire series of general knowledge questions such as, 'Can you name a precious gem?' and had to complete simple timed tasks such as identifying a pattern. Participants rated by their friends as highly charismatic were faster on the mental speed tasks, even after taking into account general intelligence and personality. Mental speed was actually more predictive of charisma than was IQ.

Charisma can be thought of as a specialist form of social intelligence, especially valuable for leaders and influencers. On the other hand, it can also be a font of emotional contagion. It can be a motor for group motivation or a weapon for manipulation that disables emotional and social intelligence in others, which helps to explain why groups often get less effective around charismatic leaders.

Uffe Schjøedt and Andreas Roepstorff at Aarhus University carried out a study that identified one possible mechanism for how this might occur at a neurological level. The focus was on brain activity in response to contact with highly charismatic people. In this case, Christian participants lay in a scanner having their brains imaged while they listened to audio files of church speakers whom they believed had healing abilities. The scan results showed that listening to the healers deactivated activity in the prefrontal cortex, the area involved in executive functions including memory, planning, reasoning and flexibility in thinking. Researchers were able to use the level of deactivation to predict the participants' rating of the speakers' charisma and experience of God's presence during prayer.

It seems that charismatic influence depends partly on a dulling down of executive functioning in the brains of the affected group. The extent to which the group believes in the power and authority of the charismatic person is also crucial. Religious belief is by no means the only channel for charismatic influence. As the study's authors point out, where charisma and authority are combined, as they are in the leaders of many different organisations from schools to youth groups to businesses and political parties, the effect on groups can be extreme – for good and ill.

None of us should assume we are immune to emotional contagion and unconscious influence by manipulative or charismatic forces. One way we can protect ourselves and boost our emotional resilience is through regular mindfulness-based meditation, which has been shown to make a measurable difference to people's ability to tune in to emotions. It also supports compassion and resistance to emotional contagion, particularly when practised in a group.

Corporations now regularly incorporate meditation into their approach to teamwork. Boston Consulting Group worked with organisations as different as Shell, Goldman Sachs and the European Commission, offering workplace meditation to create a culture of emotional safety and mutual support. Individuals were offered ten-week meditation courses and groups were then encouraged to meditate together every day. As little as two minutes at the beginning of a team meeting was shown to promote active listening and greater cooperation. Checking in and checking out of meetings with short personal enquiries about how participants were feeling also contributed to a sense of being valued and heard. Taken together, these sorts of measures seem to boost our emotional intelligence and offer some protection against unconscious influencing.

Complexity is a by-product of boosting our collective intelligence, but it is not beyond us to train our minds to cope with it. There are many strategies we can adopt to support our resilience, patience and empathy. One of the most helpful of all, perhaps, is to remember that cognitive friction is not a sign of failure but a signal that we're collectively doing something right. Joined-up thinking is a habit we can all cultivate.

EXERCISE: Cultivate Curiosity Rather than Fear

If you're feeling overwhelmed or stressed, try imagining how different brain regions are involved in the way you think. The fear response arises from the amygdala, the small almond-shaped structure in the middle of your brain, whereas calmer, more rational thinking arises from the prefrontal cortex, the large region behind your forehead. Imagine these structures in your brain as the focus for five minutes of calm.

Now set a timer and sit in a comfortable position. Close your eyes and bring to mind a situation you've been involved with recently that felt unsettling. Nothing traumatic; perhaps a sharp word with your partner or a disagreement with a colleague. Now, take some deep, slow breaths in and out. Allow yourself to replay the scenario and feel the feelings of irritation, threat or uncertainty that come up. You might need a minute or two to let them subside a little. Then, picture yourself responding differently from the way you did at the time. If you dismissed your colleague's point of view, can you now cultivate some curiosity about why they might think that way? What could you learn? If they dismissed yours, can you try to imagine what was motivating them, exercising as much empathy as possible. Imagine your small, almond-shaped amygdala shrinking even further and dimming in activity and your prefrontal cortex, behind your forehead, lighting up with activity, forming new connections in your brain as you explore different responses. Focus your attention on the idea of a brain that's busy learning from interactions rather than closing down in response to them.

CHAPTER SIX

Intuitive Intelligence: The Next Great Untapped Skill

We probably all know somebody we would describe as intuitive. They're the emotionally intelligent friend who just seems to *get* people, or the wise colleague who can solve complex conundrums and sail through the day whatever is thrown at them. Intuition can feel mysterious. Is it an actual skill or is it a hard-to-define personality trait? Is it merely a coincidence or something we only identify in hindsight – no more than the result of our brains' ingrained preference for spotting patterns in random data? And if intuition really does exist, how does it tie in to intelligence?

It was the pioneering twentieth-century psychologist Carl Jung who coined the term 'collective intelligence'. Central to his work was the phenomenon of intuition, which he regarded as the ability to tap into a collective consciousness. For Jung, intuition was fundamental to perception, which, as we know, is the foundation for the brain's capacity to construct a sense of reality.

Jung's ideas went through a long period in the intellectual wilderness, consigned to the realms of mysticism. Lately though, there has been a resurgence of interest in some areas of his work. In 2017, journalist Bruce Kasanoff wrote an article for *Forbes* that has been read more than 1.3 million times entitled 'Intuition Is the Highest Form of Intelligence'. He described intuition as 'a clear understanding of collective intelligence . . . . most websites are today organised in an intuitive way, which means they are easy for most people to understand and navigate. This approach evolved after many years of chaos online, as a common wisdom emerged over what information was superfluous and what was essential.'

This idea that something should be intuitive to use feels obvious, but we don't necessarily stop to wonder what skills intuition draws on. Does it rely on a fine-grained understanding of other people's behaviour or emotions? Does it originate from different sensitivity and messaging within the so-called 'second brain' of the stomach? If the neural underpinnings to intuition have been discovered, would it benefit us to learn how to tap into our abilities?

In the last few years, cognitive scientists have been investigating these sorts of questions and coming up with fascinating answers. The new science of introspection shows us that hearts and guts are crucial elements in our cognitive organism, that some people really are super-sensitive, and sheds light on how ideas are spread unconsciously from person to person. The implications of intuitive intelligence are profound in so many areas of our lives, from our health, bank balance and relationships to the success of our groups. But what exactly is happening in the body and brain when we intuit something, and how can we measure it, if it's happening unconsciously?

Intuition Is a Superpower

Intuition is linked to our interoceptive ability, which begins with our bodies' enteric nervous system. This connects the heart to the gut, brain and immune system and monitors internal functions such as heart rate, digestion and temperature control in order to construct a moment-by-moment map of the body's landscape. This can then be used to direct emotions at a conscious or unconscious level and so alter our behaviour, moving us out of harm's way or into a cooler or warmer environment.

Intuition is a complex and largely unconscious whole-organism function that we are only now starting to understand. As pioneering interoception scientist Joel Pearson of the University of New South Wales says, our intuitive intelligence is driven less by quasi-mystical moments of insight and more by a largely unconscious process of sifting through vast amounts of information and extracting what's useful from what's irrelevant. This of course is what our brains do all the time anyway, but our embodied cognitive system can supplement the brain's constant work.

The need for intuition arises from the fact that there's just too much information coming at the brain for it to handle. As we've seen, much of this information is simply lost. Other bits are processed without conscious awareness but registered as an emotion, or as a gut feeling, a heart-flicker or a hunch.

It can be a stretch to grasp the reality of the unconscious realm but it might help to consider the case of Patient DB, which was documented by psychologist Lawrence Weiskrantz back in 1974. This patient was absolutely convinced that he was blind on one side of his visual field and yet he was able to detect and sort between various stimuli that were presented to him there, to a far greater extent than could be explained by chance alone. He had a condition that Weiskrantz called 'blindsight'. His brain was able to perceive and process information without his conscious awareness.

We all operate in this way, even if our vision is completely unimpaired and our minds free of any delusions. The unconscious mind is constantly assessing information and making decisions that the conscious mind subsequently assumes it made all by itself. We feel the result of this on an emotional rather than at a rational level and it's this phenomenon that's increasingly thought to underlie intuition.

Some people seem more aware of their intuitive responses than others. These 'super-sensors' are better able to access their gut instincts fluidly and frequently, and better able to trust their hunches. This can make them more resilient in situations of doubt and uncertainty and less prone to negativity bias, which is the brain's universal tendency to assume the worst – a potentially useful safety mechanism that can also block our ability to try new things or take worthwhile risks. The advantage to decision-making can be huge, and can lead to very tangible results.

When Jack Soll, professor of business at Duke University in North Carolina, studied economists' forecasts and their success rate, he found that the top forecaster was about 5 per cent more accurate than the average. So even in a group of experts, whose skills had been built up from years of experience, there were those who were significantly better than others. What was the explanation for this marked ability in some individuals? Where was it coming from?

Former trader turned cognitive scientist John Coates had a hunch he knew, so he set out to investigate. Coates used to work as a financial trader on Wall Street and, while making huge profits, he started wondering why some traders were better than others. Did they have quicker computational brain power? Were they capable of making better, faster decisions?

John assessed himself and those around him, and thought a different explanation might be more likely. 'Traders often speak of the importance of gut feelings for choosing profitable trades – they select from a range of possible trades the one that just "feels right",' he told me. He wondered if introspective ability was involved, and his hunch led him to take a sabbatical from finance and retrain in neuroscience and physiology at Cambridge University, which is where I met him in 2016.

What he found, admittedly in a small study analysing just eighteen traders compared with forty-eight high-functioning non-trader controls (aka university students), was that the financial traders were far better at reading their 'gut feelings' than the controls. Coates measured the participants' introspective abilities using heart-rate detection tasks in which he asked them to count their own heart rate. The traders performed significantly better than the controls. Strikingly, even within the group of traders, those who were better at the heart-rate detection tasks also performed better at trading, generating greater profits. Even more strikingly, an individual's interoceptive ability could be used *to predict* whether they would survive longer-term in the financial markets when an economic blip occurred later that year. Coates told me that the most successful traders 'manage to read real and valuable physiological trading signals, even if they are unaware they are doing so'.

Intuitive intelligence seems to significantly boost real-world performance and add an extra strand of skill to the crystallised intelligence built up through expertise. It also seems to increase when we are exposed to other people. 'Interoceptive cues absolutely come from social groups,' John informed me. 'It's difficult to trade on your own, for example. You get tons of information without knowing it just by being on a trading floor, even if you aren't getting explicit trade ideas from other people.' John told me that many lone traders will actually pay for a spot on the trading floor in London or New York, so great is their conviction that they need to be there in person, among the crowd. If his study is to be believed, they are right to trust their gut instinct on this.

What type of information might super-sensors be picking up on when they're reading the mood in the meeting or on the trading floor? The answer is, a whole load of non-verbal communication. It's long been known that the body gives off signals that can unconsciously steer both the individual and the collective. The smell of fear, for example, where we secrete chemical signatures in our sweat that subconsciously warn others of a potentially dangerous situation, or the look of love, where a woman's pupils hugely dilate when she is at her most fertile and looking at someone to whom she's attracted. We all use this information to give colour and detail to our understanding of the world, without ever being aware that we're doing so. It seems that sharing of skills and knowledge doesn't only take place at a conscious level, via language, but also unconsciously, via the millions of bytes of data that our brains can't process – but our bodies can.

Updating the Wisdom of Crowds

People have known for hundreds of years that – at least in some circumstances – a crowd can display greater wisdom than the sum of its parts. 'The wisdom of crowds' is the suggestion, first articulated by Francis Galton, eighteenth-century intelligence researcher and all-round polymath, that if you asked a crowd of people for their best answer to a question and averaged their guesses it would turn out to be accurate. The example he stumbled across involved a crowd of 787 people at a country fair who took part in a competition to guess the weight of an ox. Nobody got the correct answer (1,198 pounds) but the average of all 787 entries was an impressively close 1,197 pounds. Galton was so struck by this that he set out to replicate his finding and wrote up his results in the new scientific periodical *Nature*, dubbing the phenomenon 'the wisdom of crowds'.

Since then it has been much studied. Jack Soll's investigations looked at a group of economists who were known for their highly accurate predictions, comparing their economic forecasts with those of their co-workers. He analysed around 28,000 forecasts and calculated the average. This problem was much more complex than the weight of the ox and drew on more background knowledge and expertise, so Jack expected the highly rated economists' average to be slightly better than the larger group's, which it was. (In much the same way, a group of farmers were more likely to arrive at a correct answer about an ox's weight than a group of city dwellers.) Soll also anticipated that the small group's average would be better than any of their individual scores, but was unprepared for just *how much* better. What he found was that the average prediction from the top bunch of experts was a whopping *15 per cent more accurate* than any of their individual predictions. Even when upping the problem complexity, the average of the elite group's predictions was always better than anything any of them had managed alone.

It's tempting to assume that crowds are only wise when it comes to matters that can be crunched down to mathematical averages or when all the individual members are experts, but that's not the case. When contestants on *Who Wants to Be a Millionaire?* opt to 'Ask the Audience', for example, the studio audience's answer is, on average, correct an impressive 91 per cent of the time.

So what's going on? Part of the explanation seems to be that crowdsourcing an answer to a problem can, if you ensure that certain conditions are met, have the effect of filtering out error while retaining insight. The question needs to be put simultaneously straight to each individual, and there should be no conferring. It needs to be the kind of question that an individual can assess using their own knowledge or from the parameters of information given, something in the general realm of knowledge such as the weight of an ox for a crowd of eighteenth-century farmers, rather than something highly specialist such as the Latin name for a certain species of orchid.

In his 2005 book *The Wisdom of Crowds*, the *New Yorker's* business columnist James Surowiecki identified four key attributes that enable the phenomenon:

1. **Independence:** Each person must guess without knowing what other people think, to stop people gravitating towards an erroneous consensus.

2. **Diversity:** The group must comprise individuals from various demographic, cultural and expertise backgrounds to ensure the contributions span different ways of thinking and do not reflect one bias that is then amplified by a clone-like team.

3. **Decentralisation:** Each person makes a guess based on their own knowledge, having the confidence to draw on their own set of expertise or experiences.

4. **Aggregation:** There must be some way of aggregating the guesses into a single collective guess, by averaging all the guesses for the weight of the ox, say, or plumping for the majority answer in *Who Wants to Be a Millionaire?*

There are some obvious limitations to using the wisdom of crowds for highly complex problems, but they don't nullify the main argument that a group *can* pool its information and cognitive resources if it behaves and communicates in a certain way. The familiar concept of the wisdom of crowds does seem to tally with the science around diversity and collective intelligence.

When Is a Hunch Just a Prejudice?

Group intelligence may depend in part on members' abilities to tap into each other's information sources unconsciously, by interpreting gestures and expressions. But if much of this reasoning is happening without our conscious awareness, how can we control it? Can we even trust it? After all, if we know that we're all subject to unconscious bias, can we really depend on our hunches or gut instincts?

We know that the greater interoceptive ability someone has – the more aware they are of their own heartbeat, thermostat and digestion – the greater their awareness of their emotional response and the higher their intuitive ability. But strikingly, this skill could either be a powerful ally, helping people make accurate decisions, or it could hinder them. If their emotional response to something in the environment is inaccurate, perhaps based on a past experience that bears little relation to their current situation, their hunch might be inaccurate and misleading. This reflects the mixed blessing that intuition confers. Sometimes, a hunch really is just a bias.

Jennifer Eberhardt, professor of psychology at Stanford University, has been employing brain imaging to demonstrate the workings of unconscious bias. 'It stems from our brain's tendency to categorize things,' she explained to *Science* magazine, 'a useful function in a world of infinite stimuli, but one that can lead to discrimination, baseless assumptions and worse, particularly in times of hurry or stress.'

Her research was inspired by her experiences growing up, firstly in a predominantly African-American suburb in Ohio before her family moved to a mostly white suburb. She noticed striking differences in the way her family was treated there, compared with the white neighbours. Her father and brothers, for example, were frequently pulled over by police for routine searches. In a recent study, Eberhardt analysed the brain activity of twenty students, ten who were white and ten who were Black, while presenting a series of photographs of faces. When students viewed faces of their own race, brain areas involved in facial recognition lit up and they had more ability to remember details later. When they looked at faces of people from a different racial background, they struggled to recall as much detail.

Eberhardt believes this is an example of learned perceptional bias that occurs in one of our earliest stages of sensory perception. Our brains' filtering mechanism for processing incoming information is based on the library of data we've built up over our lifetimes. When it comes to perceiving faces, the racial diversity of the environment we grew up in will affect our skilfulness. If you grew up among white people, you learned to make fine distinctions among whites. It's not a biologically ingrained race blindness, but the effect of environmental fine-tuning.

Experiences of bias can negatively affect a person's decisions and choices in any number of ways. They can also have immovably tragic consequences. The probability of being shot by the police when unarmed and Black is 3.49 times higher than if you are a white, unarmed person, according to an extensive analysis of police shootings in America between 2011 and 2014 conducted by Dr Cody Ross of the Max Planck Institute for Evolutionary Anthropology in Leipzig, Germany.

These sorts of startling statistics motivated Eberhardt to investigate how perceptual tuning might lead to these terrible outcomes. She used a technique called a dot probe paradigm, which involves asking volunteers to stare at a dot on a computer screen while subliminally implanting images into their minds. A group of white people including police officers and students took part, staring at the dot as Eberhardt flashed a series of images at them imperceptibly quickly. The images consisted of either a black face, a white face, or no face at all, interspersed with a vague outline of an object that gradually came into focus. This object could be a threatening one such as a gun or knife, or benign, such as a pair of sunglasses. The subjects were asked to press a key as soon as they recognised the object.

When primed subliminally with a Black person's face, the subjects recognised the weapon significantly more quickly than participants who had seen white faces. The researchers reversed the experiment and found the same thing: flashing subliminal images of weapons followed by a brief image of a face. This time the unconscious image of a gun primed the brain to identify a black face more quickly, but did not affect identifying a white face.

Racial bias was ingrained in the study's participants: the concepts of a black face and a weapon were indisputably linked. Eberhardt had demonstrated that negative racial stereotypes tend to associate Black people with threat, and this often leads to the misidentification of harmless objects as weapons when they are held by a Black individual.

Even the phase of our heartbeats – our blood-pressure reading – can affect how we subconsciously process information and make inferences from it. Sarah Garfinkel, professor at University College London, built on Jennifer Eberhardt's investigations. Sarah observed that activation of race-threat stereotypes synchronised with people's heartbeats, and that there were more race-driven misidentifications of objects as weapons during systole, which is the phase of the heartbeat when the heart muscle contracts and pumps blood from the chambers into the arteries. This is when pressure firing into the brain from the heart is at its maximum. It's a brief rise in blood pressure.

This incredible finding ties very specific heartbeat cues to threat appraisal and links them to racially biased decision-making. Garfinkel concluded in her report that since anxiety and stress send our heart rates higher and can 'exacerbate intergroup biases, these results will be of potentially crucial importance to the training of police officers and health workers'.

But the implications don't stop there. Blood pressure is rising in the general population, linked to stress, dietary changes, sedentary lifestyles and the side effects of prescribed medications. Does this feed into rising rates of fearfulness? The knock-on effects of Garfinkel's findings raise profound questions about how society might need to mitigate these fear-based interactions and the biased treatment of individuals that can result.

The impact of fear on our own brain health can also be marked. Garfinkel's team has found that the greater our awareness of our bodies' internal functions, the higher our intuitive abilities and the more sensitive our fear response. Measuring people's heart rates in response to threat shows that increased heart rate can intensify feelings of fear and anxiety but can also block out pain, making our fight-or-flight responses easier to carry out. The brain region involved in these fear responses is the insula, which is generally associated with emotional regulation, and it seems that the size and sensitivity of the insula correlates with an individual's ability to detect their own heartbeat.

There are clear positives and negatives here. On the considerable downside: if our interoceptive awareness is oversensitive or misfiring, the brain is liable to overreact, or incorrectly register, any signals that alarm us. This might be one of the biological triggers for depression, anxiety, eating disorders and autism. The upside is that people who have the ability to tune in to their intuitive, embodied cognitive intelligence are able to pick up on a lot of that extra information that is registered unconsciously, in their hearts and guts. This seems to give them a considerable advantage, especially in terms of social and emotional intelligence in situations where they're required to read the room, get a sense for how other people are feeling and make decisions accordingly.

Particularly intuitive people seem to have finely tuned introspective systems that are sensitive but not overly so. They can benefit from the extra information that their embodied cognition is picking up on, without overreacting and misinterpreting it.

Can We Learn to Be Super-sensors?

Is there any room to bring conscious awareness to our introspective functioning in order to pick up on more of these incoming signals? Can we train ourselves to dial down our fear responses so that we can put more trust in our gut instincts? Joel Pearson, director of the Future Minds Lab at the University of New South Wales, Australia, and his team set out to investigate.

In 2016, they published a study claiming that they had found a way to measure the impact of introspective functions on intuitive decision-making. Their experiment depended on a similar technique to the one used in Eberhardt's investigation into unconscious bias. They recruited a small group of around twenty volunteers, presented each with a snowstorm of dots on a computer screen and asked them to record if the dots were moving left or right. At the same time, and without their knowledge, the volunteers were presented with emotional images that had been rendered invisible through continuous flash suppression, a method first described by neuroscientist Christof Koch in 2004. One eye is presented with a static image while the second eye is presented with a series of rapidly changing images.

The results showed that after a priming period, when the participants had the subliminal images piped into their brains they were around 10 per cent more accurate in determining which way the dots were moving; *and* they responded more quickly *and* reported feeling more confident in their choice. This improvement didn't occur when the volunteers were not presented with the subliminal image.

Joel is interested in how the body rather than the brain responds to such subliminal information, and how this might differ between individuals. One way to measure this type of subconscious reaction is by assessing electrodermal activity – the skin's tendency to increase sweat-gland activity when a person is exposed to certain stimuli.

When your brain receives signals that emotionally arouse you, challenge or surprise you – making you feel scared or angry, say – then the autonomic nervous system kicks in to regulate your breathing, heart rate and arousal state. This triggers your sweat glands, which in turn increase the skin's ability to conduct electricity, a change that can be picked up by electrodes attached to the skin. All of this happens unconsciously in response to any number of stimuli, from strong emotions to unexpected events or a difficult intellectual task.

Electrodermal response tests (EDA) underpin the so-called lie-detector tests that have been used in the US legal system, but they can't indicate much about the specifics of what stimulated the reaction. They are nonetheless a useful tool for determining that *something* is happening in the unconscious.

First up, in Joel's study the EDA result was used to infer how good a person was at picking up internal subliminal cues that in turn help us to make good 'intuitive' decisions: so the EDA acted as a marker for intuition. The experiments also suggested that participants became better at accessing and using their intuition over time, steadily improving their decision-making by around 5 per cent over three blocks of trials.

Joel is pretty excited about this part of the result, since. '[It could be used] to train people to rely more on the emotional information in their brain and their body, rather than just logical, conscious information. In the future, we might be able to develop a method to train people to take advantage of their intuition and then test them to see if their intuition improved with practice.'

Sarah Garfinkel's work also suggests there may be ways to improve our interoceptive ability through training. In 2021, she trialled a pioneering therapy aimed at lowering anxiety for people with autism by getting them to tune in to their own heartbeats. This might offer a valuable new therapy for people who, for example, experience so much anxiety in groups that it stops them from being able to participate fully in social life. That anxiety can be felt as an unbearable distraction that prevents people from being able to think clearly or interact calmly. One participant in Sarah's trial was enthusiastic about the benefits. She described the positive link between strengthening her awareness of her heartbeat and lowering her sensitivity to stimulus coming from outside. 'As the inner channel gets clearer, the outer channel gets quieter.'

Sarah suggests that it could be beneficial for any of us who feel that we are overwhelmed with too much external or internal noise. If you want to try it, her protocol for testing and exercising your own interoceptive awareness is coming up at the end of the chapter.

Does Intuitive Intelligence Work Remotely?

What happens to intuitive intelligence if we can't access all the subliminal information our colleagues are giving off because we only ever see them in meetings over Zoom? At an even more basic level, what happens to chance encounters with colleagues from different departments that spark a useful thought? What happens to the conversations that drive all intelligent group behaviour, to the social bonding and negotiation of power plays? Can any of us, however sensitive our introspective functioning, however diverse our team, really work together intelligently as a group when we're only connected remotely?

Those of us who worked from home during the Covid-required lockdowns will probably have an opinion on this. For some of us, remote working was a source of frustration. We realised how much we depended on the interactions we had with people in the office. There definitely was a measurable cognitive decline for large numbers of people during lockdown, one that remote-working technology was not enough to offset.

Other people blossomed. They found the quiet of working from home a relief from the distracting bustle of the office. They were more focused and just as productive, if not more so. Could it be that extroverts struggled and introverts flourished? Perhaps individuals' work on a project was relatively unaffected but aggregating their outputs into a single presentation, say, was made harder by remote working?

The scientific evidence is mixed. Miguel Nicolelis is professor of biomedical engineering at Duke University. He and his colleagues performed some fascinating research into how monkeys collaborate in pairs to complete a task, watching how their brain activity synchronises as they work together. Pools of neurons in both individuals' brains start to fire at the same time when the monkeys tackle a task, and synchronicity increases as the physical distance between the two monkeys gets smaller.

Dr Vicky Leong carried out the work on the synchronisation of brain waves and its impact on our ability to communicate that we looked at in relation to how babies learn, back in Chapter Three. Perhaps her findings help to explain why the monkeys' problem-solving abilities went up the closer they were to one another?

Leong discovered that direct eye contact between individuals has an astoundingly positive effect on information transfer and rate of learning. It is a simple and powerful way to bring about the synchronisation of brain waves that supports language acquisition and socialisation in babies, and bonding between babies and caregivers. But . . . only for in-person communication.

Communication via a screen is not sufficient. It's the live feedback loop of synchronised brain waves, transmitted by direct eye-gazing, that makes the difference.

So does this mean that remote working is generally speaking a drag on intelligent group work, no matter the individuals' preferences? Maybe not. In May 2021 Professor Anita Woolley and her collaborators, who included Professor Thomas Malone from the MIT Centre for Collective Intelligence, published a meta-analysis of twenty-two different studies into group performance. She and her colleagues were hoping they would be able to come up with a collective intelligence quotient or CI, a way to measure group intelligence that was as reliable as the tests that measure an individual's intelligence quotient, or IQ. Strikingly, Woolley and Malone found that collective intelligence was not significantly affected by remote working. It didn't seem to make much difference whether the group was in the same physical space or gathering online.

The studies included data on more than 5,000 individuals in more than 1,300 groups. They were from various sectors including military personnel, online gamers and university students. They worked in different settings, including face-to-face versus online, and in groups of acquaintances versus groups of strangers. There was a range of tasks, from generating as many ideas as possible to finding the one correct solution to a problem and executing a specific task as accurately and quickly as possible.

When all this varied data had been crunched, the researchers concluded that the proportion of women in the group continued to be the single most relevant predictive factor, and that, as before, social perceptiveness was the explanation. A group where social perceptiveness was both reasonably high and broadly similar across the group had a higher collective intelligence quotient. Intriguingly, high levels of age diversity had negative effects on the group's collective intelligence quotient. I have a hunch that this might speak to that expertise bias we saw in the *New Scientist* readers who couldn't recognise that they had limitations, though of course my intuitive intelligence might be off here!

This finding about the lack of impact of remote working surprised me, but perhaps it speaks to our species' incredible capacity to adapt to our environment? In which case, the wholesale shift to Zoom meetings during the Covid pandemic lockdowns will have given a turbo-boost to this story of increasing fluency at group work.

Might the pandemic turn out to have been another shove for humanity along the evolutionary path away from individuality towards becoming a cognitive superorganism? More and more data is becoming available that demonstrates how good we are at adapting to our new environment and how we can still pick up on information without needing to be present in person. In April 2022, an interoceptive study was published showing that we can correctly assign the audio recording of a heartbeat to the right individual simply by looking at a video of them, without the person talking or moving other than blinking. This is one of those findings that makes me marvel at the extent of our untapped intuitive intelligence.

The data on collective intelligence is itself evolving fast. If we are entering a period of evolutionary shift, our powers of intuition and our interoceptive ability will certainly have a big part to play.

EXERCISE: Training Your Introspective Ability

Find a quiet place and make yourself comfortable.

Set a timer for one minute.

Close your eyes and take some deep breaths as you tune into your body. You're going to try to detect your heartbeat simply by listening to or feeling its vibrations. This can be tricky, so be patient.

When you're ready, start the timer and count your heartbeats for a minute.

Repeat the exercise, this time taking your pulse in your wrist or neck. Observe how much easier this is and whether your previous assessment was accurate.

Tip: if you're struggling to detect your heartbeat, try running up and down some stairs or doing star jumps until your heart pounds. Then try to stick with it as it returns to normal.

CHAPTER SEVEN

The Hive Mind and Humanity's Dark Side

Which creature on Earth causes our species the most harm? Sharks are pretty bloodthirsty, saltwater crocodiles aren't to be snuggled and a mosquito's buzz can herald the arrival of a toxic gift to our bloodstream. But what tops the list as the deadliest species to humanity? Unfortunately, that would be us. We've been killing our own kind for around 10,000 years. War alone is estimated to have caused between 150 million and 1 billion deaths. Man-made climate change and pollution, the destruction of the natural world and associated global pandemics, not to mention our consumption of vast quantities of toxins linked to cancer, diabetes, strokes and obesity, are all sending this high number even higher. In this light, the trajectory for humanity looks grim. Which rather begs the question, is there really such a thing as collective intelligence? Collective stupidity, more like.

It's time to dive into a truth that we would be foolish to ignore: sometimes, groups encourage themselves to behave in self-serving and limiting ways. For every example of collective triumph there is another of collective apathy, failure of imagination or flawed thinking. Sometimes the results are merely self-defeating; sometimes they are tragic. History is littered with incidents of genocide, warfare and the oppression of countless indigenous people. Tragically for our species, there are far too many examples of atrocious collective thinking and actions.

I want to look at how a sort of collective stupidity emerges, so that we can recognise it and steer in a different direction. Long before we get to genocide there is a range of behaviours that lead to apathy, polarisation of opinion, tribalism, complicity, manipulation and oppression. These behaviours are sometimes pragmatic, if not necessarily admirable or smart, in that they serve the short-term interests of an individual or their immediate group. Sometimes, though, they don't even have that to recommend them. They derive from our blind spots, habitual thinking, bias, or our susceptibility to manipulation and the pleasures and comfort of belonging to a tribe – however deluded or criminal.

But when we know more about how these negative group dynamics operate, we can choose to guard against them. We can train ourselves to resist the demands of instant gratification when we're making decisions, and instead use the prism of wider benefits for the group and longer-term benefits for ourselves. We can design and implement systems that nudge our decision-making in directions that benefit us all. We can build our immunity to contagion and conformity through understanding that we are all much less rational than we realise. We are all vulnerable to forces that seek to manipulate, divide and disempower us. Our species is hard-wired to care very much what the group thinks of us, which means that few of us are immune to our need to fit in. Forewarned about the power of the hive mind truly is forearmed.

Contagion: How Groups Get Infected with Bad Ideas

Human beings turn out to be a lot like apples. Just as 'one rotten apple spoils the barrel', so one cheating participant has been shown to turn a group towards cheating. Behaviours, emotions, ideas and values can spread through a group in ways that recall the ethylene gas and mould spores that are responsible for that rotten apple.

Groups of people synchronise their emotions in various ways. For example, if somebody smiles at us, we are likely to smile back and to experience a lift in our mood as a result. Sadly, the very same impulse to mimic contributes to waves of suicides that can spread through a particular group. The factors in such situations are way more complex than they are when we return a smile, but the spread occurs in similar ways.

Emotions are particularly susceptible to contagion because they key into unconscious thinking. Once an emotion has gone viral, it will influence conscious thoughts and then actions across the group. This spread has been observed in direct contact between pairs of individuals, in smaller groups and larger crowds. It doesn't need to be 'in real life'; it works very effectively online. And, as we've already seen, it doesn't need to be a sincere emotion in order to be contagious. Anybody who has ever been moved to tears by a tragic moment in the cinema or at the theatre knows that even when events are staged we are still liable to be swept up in the manufactured emotion as it ripples through the auditorium. Studies have shown that even exposure to emojis on a text message can alter somebody's emotional state.

The effects of emotional contagion are not always as fleeting as our tears in the cinema. The World Health Organization emphasises happiness as a component of health, partly due to its protective effect against mental illness but also because it seems that we can 'catch' happiness as we 'catch' the smoking habit or the overeating habit. Studies show that obesity and smoking, which are both heavily influenced by the social norms of an individual's groups, can be understood as contagious diseases. The Framingham Heart Study, which began in 1948, suggests that the same model applies to mood: an individual's happiness depends on the happiness of others with whom they are connected.

One of the study's most intriguing suggestions is that 'clusters of happiness result from the spread of happiness and not just a tendency for happy people to associate with similar individuals'.

So having a 'friend who lives within a mile and becomes happy increases the probability that [you] are happy by 25 per cent'. A spouse becoming happy, by contrast, only boosts the probability that you will become happy by 8 per cent. This counterintuitive finding suggests that we might be able to develop immunity to emotional contagion through repeated exposure, and build it consciously by careful exposure to small doses. We will come back to this idea later in the chapter.

These results highlight something crucial. We don't fully understand how emotions are transferred. We can observe that they are, but we struggle to say for certain how it happens. The authors of the Framingham Heart Study were upfront about this, admitting that they were unable 'to identify the actual causal mechanisms of the spread of happiness but various mechanisms are possible. Happy people might share their good fortune (for example, by being pragmatically helpful or financially generous to others), or change their behaviour towards others (by being nicer or less hostile), or merely exude an emotion that is genuinely contagious (albeit over a longer time frame than previous psychological work has indicated).'

Some scientists believe the answer lies in mirror neurons, a subset of neurons that make up about 10 per cent of the total in certain brain regions. They were discovered during observational studies of macaques' brains and behaviour carried out by Professor Giacomo Rizzolatti in 1992. The researchers noticed that certain brain cells were activated both when a monkey performed an action and when that monkey watched another monkey perform the same action.

Mirror neurons are involved in learning through imitation, firing up, for example, in babies' brains as the child mimics its carer's expressions and copies their gestures. They are also implicated in our ability to imagine what other people are experiencing. So when you wince as your partner stubs their toe, a phenomenon called 'neural resonance', that's mirror neurons at work.

This act of imagining extends to being able to empathise with somebody's emotional pain when we see them crying, or talk to them about their distress.

Mirror neurons also seem to be involved in emotional synchronisation, which may be one of the mechanisms by which contagion occurs. The positive side of emotional synchronisation is that it can support healthy social development and build the core skills of emotional intelligence such as empathy and consensus-building. On the dark side, it may also be the mechanism that allows fear and hostility to spread.

Tanja Wingenbach at the University of Bath published a study in 2020 that described how people mimic emotional expressions even when they're exposed incredibly briefly to them. When participants witnessed a facial expression for just one second – so certainly not long enough to form a conscious bond with the individual – there was an increase in electrical activity in the facial muscles needed to mimic them. This result of mirroring was shown across a variety of distinct emotions, ranging from sadness, surprise, happiness or pride to anger, fear or disgust.

Interestingly, there is also evidence that this process is linked to social status. In Professor Rizzolatti's original study, lower-ranked monkeys frequently imitated a monkey of higher rank but the more dominant monkey was unlikely to be strongly influenced by the more subordinate. This finding might well translate to human beings. As we've already seen, charismatic people are generally more effective transmitters of emotional contagion, which gives them emotional sway over those in their orbit. It may be that their charisma makes them more likely to end up in positions of influence and contributes to their higher status in a group.

Emotions are coming under more and more scientific scrutiny. Psychologists have always treated them as a source of useful information about ourselves and others, while also highlighting that they are no more trustworthy than any of our other expressions, in words or actions. Recent studies have built on this to break down the idea that emotions are a kind of universal language that transcends cultures, and can be reliably read from your face. Lisa Feldman Barrett at Northeastern University and Carlos Crivelli at De Montfort University, Leicester have both concluded that our facial movements are not always 'true' pictures of what we feel inside but function more like tools to help us communicate and interact with others. The movements people use to express emotions change according to situation, person and the cultural context.

Barrett and Crivelli's work involves taking a pile of photographs of people posing stereotyped, Western expressions (e.g., scowling in anger, frowning in sadness, widening eyes and gasping in fear) to show to groups who live in small-scale, remote settlements. It turns out that people who live in these locations, further from the influence of Western culture, do not perceive the same emotion in these faces as Westerners do. This and other research suggests that there is no such thing as a universal set of emotional facial signatures. We perceive feelings such as threat, aggression, fear and disgust according to codes we learned from the faces we were surrounded by as we grew up, and the beliefs about emotions that we have acquired. The facial movements used to express emotions vary across the world, just like language and dialect; and, crucially, they can be used to influence, and manipulate, other people's responses. The Trobriand Islanders, for instance, don't believe that facial movements are transparent reflections of sincere feelings. They believe you 'have a face' to get what you want or avoid what you don't. This may sound obvious, but the concept that we can reliably tell what someone is feeling from looking at their facial movement underpins all sorts of applications in schools and the justice system, as well as tests for autism and even emotion-detecting-software algorithms.

The ongoing re-evaluation of how we understand the role and purpose of emotions has big implications for how we approach emotional intelligence. I don't want to suggest that all emotions are inherently unreliable. We can clearly learn a lot by tuning in to our emotions and those of other people, but we must do so intelligently, applying our learned experience and feeling around for nuance. Emotions are partly social constructs and tools for behaviour change. We can learn how to feel and express them. We can fake our own, and manipulate them in other people.

This knowledge can make us smarter and more powerful if, for example, we consciously decide to fake our confidence until we make it part of our repertoire. It can also remind us that some people might have an agenda that undercuts the interests of the group. Knowing that emotions can be faked, exaggerated and transmitted need not make us cynical but it does underline that being sentimental or naive about them is unwise.

When it comes to contagion, it's not just emotions that can undercut collective intelligence. Many behaviours that impact how well a group functions are contagious. Take apathy. You might have heard of 'the bystander effect'. This widely studied phenomenon describes the way that people tend not to step in to intervene when they see bullying or a mugging, for example, if there are other people witnessing the event who are also not intervening. Studies have shown that if a person is on their own when they encounter such a situation, they are much more likely to get involved. If there are lots of people around, the sense of responsibility seems to dissipate.

If a group is apathetic or disengaged, that can turn to complicity. When individuals relinquish responsibility, allowing themselves to be led or excusing themselves by reasoning that somebody else will act, their apathy tends to spread and become self-reinforcing. In this way the silent majority can end up tolerating decisions and actions that they do not actively endorse or that they know are not in their best interests. These forces work directly against group cohesion, undermine the possibility of a consensus emerging and can even result in the collapse of the group. Or worse.

Nelson Mandela may have reserved the main force of his condemnation of apartheid for those who designed, governed, policed and prosecuted it, but he didn't let anyone off the hook. He said that people who chose not to see and oppose the realities of the system were complicit in its brutalities. But how do wilful blindness and apathy take hold of such a huge group – of a whole society?

Power and Conformity: Flip-sides of the Same Coin

It is an uncomfortable truth about our species that most of us are highly vulnerable to conformity – we will go along with the crowd. In addition, a few of us are highly motivated by power. Put these two things together and what you get (sometimes, though not always) is the toxic combination of a leader who craves power but is losing the ability to exercise it responsibly, and a crowd of followers whose apathy and need to belong make them vulnerable to the leader's directions. Add in the evidence that leadership is associated with charisma and prowess in manipulation through emotional contagion, and you begin to see how certain power dynamics within groups will undermine collective intelligence.

What happens in a leader's brain that might make them more inclined to antisocial behaviours as they work their way up a hierarchy? And what happens in a follower's brain that makes it more likely they will be influenced in potentially antisocial ways by their leader?

It used to be thought that, quite simply, power corrupts and social pressure causes groups to collapse into mobs. In the notorious Stanford prison experiment of 1971, twenty-four male volunteers were assigned, by the flip of a coin, to be either 'guards' or 'prisoners' in a mock prison. After six days of the anticipated two-week experiment, it was called off. The 'prison guards' had embraced their sadistic roles, forcing prisoners to sleep naked, removing mattresses from their cells as punishment and putting one prisoner in solitary confinement while inmates were coerced into repeatedly banging on his door. Many participants were emotionally traumatised by the experience. The experiment's designer, Philip Zimbardo, described it as 'a cautionary tale of what might happen to any of us if we underestimate the extent to which the power of social roles and external pressures can influence our actions'.

But there's an intriguingly different way to interpret the results. Zimbardo, who was professor of psychology at Stanford University at the time, both initiated and ran the experiment and had a key role as superintendent in the prison, encouraging the unhealthy power dynamic to develop. His active participation in the role play looks, in hindsight, like a textbook case of emotional contagion rippling through a group from a charismatic leader, not to mention the very opposite of objective observational science.

So what can we reliably say about the effect of power on decision-making, empathy and other crucial behaviours that either facilitate collective intelligence, or don't?

First up, a number of studies demonstrate that for practically anybody, even a saint, simply being in a position of power increases egocentricity. Adam Galinsky and his team from Northwestern University carried out a series of neat experiments to show this. Galinsky and his colleagues wanted to test the idea that people in power struggle to recognise and adjust for others not having access to the knowledge that they do. They began by priming half the volunteers to feel powerful by asking them to recall in as much detail as possible a time when they had been in charge of other people, such as chairing an interview panel or refereeing a football match. The other half of the group were asked to recall a situation in which they had felt powerless. The volunteers were then asked to think through a number of different scenarios and to guess at a third party's response to what had happened.

In one scenario, participants were asked to imagine that they and a colleague had been to a fancy restaurant that had been recommended by the colleague's friend, where they had a particularly poor dining experience. The next day, the colleague sent an email to the friend saying, 'About the restaurant, it was marvellous, just marvellous'. Participants were asked to respond to the question, 'How do you think the colleague's friend will interpret the comment?' Responses were marked on a scale anchored between very sarcastic (1) and very sincere (6). There was no further information in the email to suggest anything other than sincerity, so no cue to presume sarcasm.

The researchers found that if participants were primed with power they were less likely to take into account that the friend did not know what they knew: that the restaurant had been awful. The power-infused participants presumed that since they themselves interpreted the comment sarcastically, a third party would too. They insufficiently adjusted their own response to account for somebody else's knowledge and perspective.

Deborah Gruenfeld of Stanford took these results a step further to show how one consequence of egocentricity is that it inclines us to see people as a means to our ends. In a series of six experiments she similarly primed volunteers into the power mode. Under these conditions, people's approach to others was more driven by whether they might be useful to them than whether they liked or esteemed them. Essentially, when Gruenfeld evoked feelings of power in people, they began to see others as objects.

Now, let's combine these results with another study, which shows that as people rise up the ranks in a hierarchy they are more likely to excuse their own moral failings even as they continue to condemn others for the same behaviour. Cameron Anderson of the University of California, Berkeley analysed archive data from over sixteen years of investigation of the effects of power in more than 11,000 individuals. He concluded that it makes people more focused on their personal goals and rewards, and better able to justify their transgressions. There have also been studies suggesting that powerful people are more likely to act impulsively, with less awareness of potential risks.

So how can we account for these changes in behaviour? Professor Dacher Keltner of the University of California, Berkeley carried out some of the work on power and impulsivity. He found that when people are primed into feeling powerful, it deactivates the vagus nerve. As we saw in Chapter Two, this is the cable of nerves that allows our brains to sense information stored in our bodies: our hearts, guts and other organs, where subliminal information from those around us might be held. It seems that power dampens down our brains' ability to process and integrate the full repertoire of information from the world around us, including the perspectives and feelings of other people.

It's admittedly early days in this field of neuroscience, but there is a growing body of evidence that explains how power tends to disable the cognitive skills on which collective intelligence depends. This tendency is just that – a general trend and not an absolute. Obviously there are inspirational, transformative people of influence out there. So, what makes some individuals immune to this corrupting force of power?

Dr Farid Youssef, lecturer at the Faculty of Medical Sciences, the University of the West Indies, St Augustine, Trinidad & Tobago, was interested in the idea that some people might have some immunity to the negative effects of power if they had well-developed social skills. He asked men and women to play the ultimatum game, a test used by behavioural scientists to investigate empathy and perspective-taking by seeing how people react to fair and unfair offers on how to split money. There was no statistical difference in the two sexes' baseline results but then participants were put through a stress test – basically having to do complex mental tests against the clock and in front of assessors – and sure enough, both men and women's cortisol levels shot up.

Then they played the game. The effects of stress *were* sex-specific: in females it triggered heightened levels of conciliatory and relationship-building strategies, whereas males were increasingly likely to reject offers, whether fair or unfair. Essentially, women opted for a 'tend-and-befriend' response and became more cooperative, whereas men went for a 'fight-or-flight' response and became less cooperative. These results tie in with a number of other studies indicating that in situations involving rewards, punishments and uncertainty, acute stress increases risk-taking in men and decreases it in women. Interestingly, it seems possible to tweak how both sexes react by altering their oxytocin or serotonin levels, or through meditation and mindfulness practice. This finding might explain the high incidence of transformational leaders being female, since women typically have well-developed social skills.

All leaders have a crucial part to play in collective intelligence. It's no exaggeration to say that a leader can send a group either way: towards self-destruction or success.

The Urge to Conform: A Risk Factor for Collective Stupidity

What about followers? Most likely all of us will recall how crucial it felt as a teenager to fit in with our peers. (I hope I will still remember it when I become the parent of a teenager.) Adolescence is a distinct period of neurodevelopment and the teenage brain is hugely dynamic, reshaping itself through rapid learning. Risk assessment is a work in progress and so is identity. Teenage and young adult years are a testing ground for social identity, as people move away from the roles they inherited from family and explore a wider range of experiences and points of view. Some of these experiences can be beneficial and playful, but cliques and bullying can also emerge and conformity to the group can sometimes feel tyrannical.

Adults often worry about or disparage teenagers for their susceptibility to peer pressure. It might be tempting to see the urge to conform as something we grow out of, but actually, it never really goes away. As we've seen, human beings are social animals. Evolution has shaped us, with good reason, to care what other people think of us. Conformity is sometimes presented as a moral failing or at least a lack of individuality, but it is a canny survival strategy and not necessarily a bad thing. It all depends on the group norms you're conforming with.

Solomon Asch, legendary Polish-American psychologist, spent fifty years conducting various groundbreaking studies into social psychology. He spent his childhood in Poland, then emigrated as a teenager to the United States, where he initially struggled to settle in, learn a new language and understand a new culture. His status as an outsider consolidated a lifelong fascination with the formation of social identity, how we conform to group norms and how we put together impressions of other people from limited information.

In a famous series of experiments, he demonstrated how easily people will alter their judgement of a situation – even against the evidence of their own eyes – to fit in with those around them. Asch asked for volunteers to participate in what he presented as an investigation into visual judgement. Each volunteer was put into a group with five to seven others, whom they believed were also participating. In fact, these people were plants. They all knew the true purpose of the experiment was to investigate conformity. Each person in the group was asked to identify, out loud, which one of three lines drawn on separate pieces of paper was the same length as the test line. The plants were instructed to nominate an obviously incorrect line. The subject of the test was then asked for their judgement, either last or in penultimate position.

Over the course of twelve trials in his (admittedly small-scale) study of around 120 male students, about 75 per cent of participants conformed at least once, going along with the suggestion that a line matched when it didn't. A small minority – approximately 4 per cent – conformed every time, twelve times in total. Most offered their own opinion to begin with but then fell in line. A little less than 25 per cent of participants never conformed, insisting (presumably with increasing bafflement) that they perceived things differently from their peers.

Most of the participants seemed unwilling to admit (or were even unaware) that they were falling in with the group until they were questioned closely, after the true purpose of the experiment had been revealed. Their explanations included that they were scared they would be ridiculed or thought peculiar, and that they wanted to fit in. These feelings reflect what psychologists call 'normative influence'.

Some participants said they believed that the group was better informed than they were: that where the difference in length of line was minimal, others could see something that they themselves could not ('informational influence'). Whatever their reasoning, the result was the same. Peer pressure is real and is not confined to adolescence. The urge to conform is so deeply ingrained in us that three out of four people will go along with demonstrably false assessments against the evidence of their own eyes.

Academic studies aside, the reason why conformity gets such a bad press is that it can have horrendous real-life consequences. 'I was just following orders' has been used throughout history as an excuse for taking terrible decisions and committing heinous crimes. It sums up the way that direction from a leader – whether explicit in the form of orders or implicit in the form of influence – combines with context and the power of conformity to push us towards taking poor or even deeply immoral decisions. Conforming to a group's codes can, in certain circumstances, lead to joining in murder.

On 13 July 1942, the German Reserve Police Battalion was stationed on the outskirts of a small Polish village. The soldiers were woken just before dawn for an emergency briefing from their commander, Wilhelm Trapp. He explained that explicit orders had arrived: the neighbouring village was home to more than 1,800 Jews who were alleged to be involved with partisans. The orders were as follows and for immediate action: round up all the Jewish men, who were to be taken to the workhouse. The women, children and elderly were to be shot on the spot. Trapp offered that if any of the men did not feel up to the task, they could step out. Around ten did so; the remaining 490 did not. They chose to carry out the massacre of hundreds of civilians. Why did more not step out when offered the opportunity?

In his book *Ordinary Men*, renowned US historian of the Holocaust Christopher Browning set out to answer this question. He trawled through accounts from the day, diaries of those involved and testimonials after the event. The battalion was not made up of zealous Nazis; they were largely working-class middle-aged family men who were not politically active. Major Trapp had offered the option to stand down with no consequences, so fear of disobeying orders was unlikely to have been a factor. (Browning uncovered proof that men who were fathers had refused to kill children on previous occasions, and had faced no consequences.) Browning's conclusion was that social pressure was the most significant explanation. Loyalty to the group and a sense of duty to support their peers trumped their aversion to committing unquestionably immoral acts.

Social pressure is more pronounced in certain contexts, such as when there is an in-group and an out-group as there was here. Even if the men of the battalion weren't idealogues, they would have been influenced to varying extents by anti-Semitic ideology. Conformity is also more of a factor where people are required to give up their individual identity to some extent, as they are in the military. Then there's the fact that, as Asch's experiments demonstrated, the ability to withstand the urge to conform is eroded by repetition. Even if you held your ground once or twice or ten times, you might not do so fifty times. The other factor is the presence of a superior giving orders, which, as Dr Emilie Caspar and her colleagues at the Netherlands Institute for Research have recently shown, is a powerful disinhibitor of both empathy and guilt.

Dr Caspar and her team set out to investigate what was going on in people's brains when they were asked to do something immoral. In a return to the central premise of the Stanford prison experiment, they wanted 'to understand why obeying orders impacts moral behaviour so much'. The researchers measured brain activity while participants inflicted pain on others. One person was assigned the role of 'agent' and the other the role of 'victim'. Agents were placed in an fMRI scanner and their brain activity was recorded while they pressed one of two buttons. The first triggered a mildly painful electric shock to the victim's hand in exchange for money. The second delivered neither a shock to the victim nor the financial reward. Across two experiment conditions, agents were either ordered which button to press or were given free choice.

The team discovered that participants tended to administer more shocks when receiving instructions, compared to when they were free to decide which action to take. The group who were given orders to administer shocks were much more likely to do so, with 50 per cent opting to obey the order at least once. The team also discovered that obeying orders reduced activity in the brain regions associated with both empathy and guilt, when compared to deciding freely. Their conclusion was that following orders dampens down the queasy after-effects that most people suffer when they inflict pain on others.

Now, the positive side of this is that when left to make a free choice, people are less likely to administer frequent shocks to others, even if that implies they miss a financial gain. When you put them in a particular situation, however, one that combines risk factors for collective stupidity, then people's innate capacity to think of others and prioritise the collective moral code is switched off. Relatively few of us – perhaps fewer than one in four – are immune to the conditions that erode our ability to think 'group' rather than 'self'.

From Psychopaths to Hyper-altruists: Ethical Thinking

So far we've been talking in general terms about how our brains process information and feelings when they're making decisions, but different brain types exist on a spectrum of sociability, with psychopaths and narcissists on one end, hyper-altruists on the other and the majority of people falling in between. As ever, the dynamics of contagion mean that when an individual with marked tendencies is in a position of influence, their values will spread through the group. This matters, because compassion and altruism make harmonious group living possible, which enables us to draw on the whole group's cognitive power and maximises our chance of ending up with collective intelligence. As we discover more about the brains of psychopaths and hyper-altruists, could this knowledge be used to tweak moral values into the future, even to bio-engineer a revolution in moral enhancement?

Altruism is by no means confined to humanity. Many other creatures on this planet have evolved with an ability to bear their own pain in order to alleviate harm for others. This is an intelligent strategy for species survival, with compassionate behaviour observed in magpies, rats, macaques and elephants among others. Altruism also has significant benefits for the individual: studies on humans have shown that doing voluntary work, for example, is linked to better physical and mental health outcomes and a longer life.

Concern for other people and an ability to put their needs ahead of our own is at the root of ethical decision-making, at least in the practical terms of everyday life. Neuroscience has recently begun to uncover how it happens in the brain, revealing the mechanisms that drive both altruistic behaviour and guilt at transgressing a personal or social code.

Molly Crockett runs a laboratory at Yale University that investigates the biology underpinning highly abstract behaviours and emotions. She and her team study the biomarkers for guilt, moral justice and social pain; the neural correlates between impatience and selfishness. Before this, she worked with Ray Dolan, Professor of Neuropsychiatry at University College London, and conducted a fascinating and robust study into compassion. The results showed that most people exhibit altruism towards complete strangers at greater levels than they do towards themselves, and even at their own expense.

This study was another to use electric shocks and financial rewards to investigate how much pain people were prepared to suffer and inflict, for what levels of financial gain. Participants had the choice to accept a lower bonus payment to reduce the number of shocks delivered to the other participant. Specifically, on each trial they chose between less money for themselves and fewer shocks for others, or more money and more shocks. The participants were largely from UCL, drawn from their international community of students. It turned out that they were prepared to pay two or three times as much money to avoid pain being given to a stranger than for themselves to avoid the same level of pain.

Ray Dolan, co-author of the study, admitted to being surprised by this. 'We made it quite explicit that they would never have to meet the person that they were sending the shocks to . . . there was no social consequence for them, no damage to their reputation. That level of altruism was totally unpredicted.'

Molly and Ray measured brain activity during the experiment and discovered that there is increased activity in the prefrontal cortex when a person is weighing up whether or not to take a financial prize at the cost of causing physical harm to someone else. This region was devaluing how the brain assessed the financial reward by dampening down the striatum, a region buried in the middle of the brain that's involved in reward perception, motivation and decision-making. It's the sensitivity of the connection between the prefrontal cortex and the striatum that underpins how we evaluate others' needs in relation to our own.

How would you feel if you lost the bicycle borrowed from your friend, which was the last present given to him by his grandmother before she died? This is an example of a question used to study interpersonal guilt. Hongbo Yu, assistant professor at the University of California, Santa Barbara, has collaborated with Molly, and is interested in what happens when we transgress our altruistic code. Yu wanted to see if he could uncover what was driving the way behaviour changes under pressure. So he induced feelings of guilt in participants by making them think they had caused unnecessary physical pain to others. Scanning people's brains with fMRI imaging revealed the regions that lit up with activity as people experienced guilt, pointing towards a neural signature for the complex emotion. The patterns were replicated across the majority of individuals from two different cultures: Chinese and Swiss. It's early days but the testing might prove relevant for identifying guilt in a legal setting.

The experiment shed light on the way our unethical behaviours seem to alter our values, and the values of other people in our groups. Ray Dolan has suggested that in order to reconcile the mismatch between our moral values, the guilt we feel by breaking them and the reward we receive by doing so, the brain will, once we have 'stepped over to the dark side', enhance the sensitivity of the striatum (priming the reward and motivation system) in order for us to feel comfortable putting our own benefits ahead of others' pain. This reset helps to build in retrospective justification for our behaviour, reduces our discomfort the next time we transgress and can ultimately send us veering off on a path of escalating selfishness.

Ray suggests that this mechanism has an impact on moral contagion. He has shown that even simple interactions allow us to pick up on clues about someone's values, often without our conscious awareness. Interactions generate a small shift in one person's values towards the other's, which seems to happen implicitly and automatically. This is one of the ways that a group builds consensus, but, as ever, it can go one of two ways. Moral contagion, like emotional contagion, can set a group on a course towards highly self-serving behaviour. Values converge around the anchor point in a group, who is typically a person with authority or influence, and often an extrovert. A person with a brain profile that inclines them to extremes of behaviour, especially if they're a leader, can turn a good group bad or vice versa.

Some people are hyper-altruists who display consistent pro-social behaviours. Pro-sociality has a genetic basis and starts to emerge early in life, increasing as children develop. The genes involved are generally those implicated in communication across the brain, especially via the dopamine system (involved in reward), serotonin (linked to mood and inhibition), oxytocin (linked to developing bonds with others) and vasopressin (involved in regulating links between the heart and brain and associated with feelings of love). People whose brains have been shaped by these particular gene variants consistently make decisions that conform with their moral code and behave more altruistically. They're 'we' people rather than 'me' people.

Psychopaths, on the other hand, have brain profiles in which the connectivity between regions involved in moral decision-making is significantly reduced. They are aware of but unmotivated by other people's feelings and points of view, and have a severely reduced capacity to experience either guilt or empathy. Psychopaths are incapable of making positive contributions to a group's intelligence, though they are often charismatic and plausible, which can make them dangerous vectors of collective stupidity.

Narcissists have a similar brain profile to psychopaths. They struggle to understand other people as independent agents rather than objects in relation to their own agenda. While psychopathy is highly heritable through genetics, and remains stable at roughly one in a hundred people in all cultures, narcissism is increasing in highly developed consumerist societies. It is currently estimated to affect five in every hundred people and American college students' self-reported narcissistic traits increased by 30 per cent in the forty years leading up to 2006. Alongside these narcissistic tendencies comes a startling sense of entitlement. In 2013, 65 per cent of students endorsed the statement, 'If I explain to a professor that I'm trying hard, he/she should increase my grade'. A third of college students also agreed with the statement, 'If I attend most of the classes, I deserve at least a B'.

Psychotherapist Dr Aline Vater from the Freie Universität Berlin wondered about the role that environment and culture were playing in this shift. She had the perfect social, economic and political context in which to investigate, right on her doorstep. Prior to reunification in 1989, West Germans spent forty years living in an individualistic culture, whereas East Germans had a more collectivistic culture. Dr Vater and her team analysed more than 1,000 people who had grown up either in the Social Democratic Republic of East Germany or in the capitalist and democratic West. What she found was that diagnoses of grandiose narcissism were significantly higher, and self-esteem much lower, in individuals who grew up in the former West Germany compared with those in the former East Germany. The difference was considerably smaller between individuals who had started primary school after reunification.

These results clearly don't show that East Germany produced 'nicer' people than West Germany, especially given that its totalitarian political regime also demanded conformity with its highly antisocial and paranoid values. They do however suggest there's some sort of link between growing up in a consciously collectivist society and behaving in ways that take the group into account.

In our own hyper-individualistic society, with narcissism at high and rising levels, an excessive focus on our self can tip into a self-sabotaging obsession. Paradoxically, it can result in dangerously low self-esteem and a fragile ego. The values of the 'We' are eroded and replaced by a self-absorbed yet highly vulnerable 'I'. That's potentially bad for the happiness and success of the individual, and has unfortunate implications for a group's ability to work together towards a common purpose and with self-restraint.

Some bioethicists have begun to make the case for bioengineering a deeper sense of morality in certain groups of people. Elvio Baccarini and Luca Malatesti of the University of Rijeka in Croatia recently argued in the *Journal of Medical Ethics* that 'the mandatory moral bioenhancement of psychopaths is justified', seeing it as a necessary 'prescription of social morality'. Treatments might include psychotropic drugs such as oxytocin to help people feel connected to others; magnetic brain stimulations for areas of the brain relating to empathy and compassion; or even neurosurgeries to help dampen down or activate specific neural circuits involved in moral decision-making.

But what about the rest of us, who aren't psychopaths and probably don't consider ourselves narcissists either? Just as we can take smart drugs to focus our analytical cognitive skills, should we be opting for procedures to boost our social intelligence? As our individual ability to contribute to collective intelligence becomes more of a currency, the more we will presumably come to value those skills. Ironically, our competitive instincts might nudge us in the direction of interventions that support our abilities to collaborate. Perhaps these treatments will eventually be as attractive as laser eye surgery, microdosing psychedelics or a serious meditation practice: just another tool that helps us to flourish.

Given our slow progress in waking up to the urgent need to

act as a collective to tackle complex problems, Julian Savulescu, professor of ethics at Oxford University, thinks we should all be stepping up for bioenhancements. He published papers in 2017 and 2012, in which he laid out the scale of the problem. According to Savulescu, we have changed our environment in such a way that our moral biology cannot keep up. We are now faced with 'moral mega-problems' and moral bioenhancement may be necessary to address them. As he puts it, 'the human predicament is now so ominous that we should not spurn developing any means that contemporary science could produce, that might significantly improve it'.

Moral science is still in its infancy and, aside from the obvious ethical quagmires, we do not have robust data to demonstrate that bioenhancements would work. Even if they did, they might also have unforeseen and undesirable consequences. There are countless examples of researchers re-evaluating discoveries as the scope of their work widens. When oxytocin was first identified, it was touted as the 'love chemical' for its ability to help create bonds and feelings of connection between people. Recently, though, it's been shown to have a dark side. It can foster empathy between a small number of individuals, but at the expense of anybody outside the immediate group. Hostility to outsiders can result.

Moral bioenhancement may remain an interesting philosophical discussion rather than a reality but the central issues it raises are valid and crucial. Given that how we value other people has a direct impact on our decisions and behaviours towards them, how big could, or should, our circle of concern be? Our innate altruism evolved to support us to live in groups of roughly 150 but when it comes to the climate and ecological emergency, our concern needs to extend not just to billions of people all over the world but to other species as well. We urgently need to rediscover our 'we' thinking and dial down our 'me' thinking.

Fortunately, the contagious nature of decision-making, emotions, values and behaviours means that when we make a concerted effort to do this, that cognitive effort is translated into real differences in our own behaviour, which then impact on other people. 'We' thinking is deeply contagious. It ripples through a group with a compounding effect, branching ever outwards. As Greta Thunberg says, 'No one is too small to make a difference'. Neuroscience is showing us the biological reality that underpins that idealistic statement, and which can sustain collective intelligence even in large groups across communities and societies.

Collective intelligence doesn't need us all to choose it at the same time for it to work. It can make its way from small pockets to larger ones as good ideas spread through our groups, generating their own rewards as they go. It can also be nudged by making positive social behaviour easier, or even the default, harnessing the influence our environment has on our decision-making. I'm optimistic about our capacity to steer towards rather than away from collective intelligence, despite the barriers we tend to put in our own way. We just need to take a clear look at those barriers, and then choose our workarounds.

EXERCISE: Be Aware of the Impact of Power

For this exercise to work, you need to perform it in real time, as you read each stage of the following instructions. Don't skip ahead!

Start by closing your eyes. You're going to recall a time when you were in a position of power. Perhaps you were a referee at a football match or chairing a meeting. Spend some time conjuring up that feeling of being in charge. Did you make decisions? Direct other people to do things? How did this make you feel?

Now, with your eyes still closed, take the hand that you write with and use your index finger to trace a capital letter E on your forehead.

Have you done it?

Open your eyes.

How did you draw the E? Would it be the right way round for other people to read it, if they were looking at you? Or did you draw it correctly for you, as if your forehead were a piece of paper you were looking at?

When Adam Galinsky and his team conducted this experiment in the lab, volunteers primed with power tended to draw an E that was correct from their (internal) point of view but appeared mirror-reversed from the point of view of someone standing opposite them. In contrast, those who were not primed with power were much more likely to think from other people's perspective and draw the E as a mirror image, so that it was immediately legible to others.

This experiment demonstrates how leaders are vulnerable to the impact of power on their brains. We can to a certain extent immunise ourselves against this effect by undertaking activities that allow us to practise our 'we' thinking: active listening and reflecting on interactions, mindfulness meditation, voluntary work and even reading fiction have all been shown to be effective.

CHAPTER EIGHT

Reshaping Our World to Boost Collective Intelligence

In the last 300 years, human beings have reshaped the environment in which we live almost out of recognition for anyone who was alive in the eighteenth century. It can be difficult to make sense of what these enormous physical, technological and social changes mean for our behaviour, but neuroscience can help us to try. Environmental pressures have a huge effect on the behaviour of all organisms, and we are no exception. Where we differ from other animals is in the extent to which we have radically altered our context, in ways that we are still only beginning to understand. Our diets, our housing, lifestyles, access to medication and, above all, our access to information via digital technology all have an impact on group dynamics and can tip us either towards or away from maximising our chances of collective intelligence. Even humanity's most brilliant and useful innovations in medicine or communication, pinnacles of collective intelligence like cheap safe painkillers or digital media, also represent barriers to our brains' full range of functioning.

We've looked at the biology of decision-making, emotions and ethics, in an attempt to understand why groups of people behave in ways that hamper their progress. Now it's time to dig into the part played by our environment. The context in which a group functions is not a backdrop to their behaviours but an active participant in them. The way we work, socialise, meet romantic partners, communicate our ideas and interact with corporations and the state is changing fast. Can we keep up, or are there signs that our cognitive capacities are struggling to avoid being overwhelmed?

Let's start by looking at some ways to engineer the physical environment we meet in, so as to maximise joined-up thinking. Then we'll look at the effects of our massive use of prescription medication and then in more detail at our even bigger use of social media and digital communication. How do these new factors in our environment affect our ability to remember, pay attention, empathise, control our emotions, evaluate information, make decisions, change our minds and listen to other people's points of view? What can we do to harness the benefits of this new world we've created while working round its unfortunate downsides?

Tweaking the Environment for Better Joined-up Thinking

It may be true that breakout areas alone will not lead to collective intelligence, but it's also true that without them, it definitely won't happen. The design of the space in which a group meets has a significant effect on how the group functions, as anyone who has ever struggled to think in a busy, open-plan office will know. As well as formal meeting rooms, it's important that chance encounters and informal meet-ups can happen easily. Smaller nooks and crannies, high-throughput corridors where people can bump into each other, canteens and outside recreation or lunch areas all facilitate different sorts of meetings. The more possible it is for members of the group to communicate with one another in easy and comfortable ways, the greater the likelihood of open communication that transcends divisions between teams and pay grades.

It's also worth ensuring that any space where we work – whether alone or in groups – is well ventilated and contains lots of air-purifying plants. Crowded and poorly ventilated classrooms, offices and air-conditioned planes and trains have all been found to have levels of CO_2 that exceed the recommended limit of 1,100ppm.

Tadj Oreszczyn, professor of energy and environment at University College London, has been researching the impact of CO_2 levels on cognitive capacity and productivity. In one small-scale study, employees' scores on an IQ test were 50 per cent lower when they were exposed to 1,400ppm of CO_2 compared with 550ppm during a working day. Our oxygen-hungry brains basically start to asphyxiate when atmospheric CO_2 levels rise too high, causing drowsiness and confusion. Oreszczyn has called for carbon-capture technology to be routinely added to ventilation systems in all new buildings and warns that the problem will only get worse as the climate crisis intensifies. In the interim, opening a window to allow a current of oxygen-rich air is the best course of action; and dotting lots of spider plants around can't hurt.

There's also the effect of temperature on brain performance. Thermal comfort is obviously important for everybody – it's hard to think when you're either very cold or much too hot. But beyond this truism, there's a small but significant difference between how men's and women's brains function according to ambient temperature. You may be familiar with the gendered battle over the central-heating thermostat or air-conditioning unit, either at home or at work. Many women prefer a warmer temperature, whereas men prefer it to be cooler, and it turns out that this preference is measurable and leads to significant differences in cognitive performance. Agne Kajackaite from the Berlin Social Science Centre and Tom Chang at the USC Marshall School of Business ran an experiment with 543 participants in which they asked a series of questions designed to test numeric, verbal and self-reflection skills, while tweaking the room temperature between 61 and 91 degrees Fahrenheit.

Just for fun, here's an example question: 'Your task is to add up five two-digit numbers (see the example below). Solve as many problems as you can. You will have 5 minutes from the time the experimenter tells you to start. You may use pen and paper, but use of a calculator is not allowed.'

EXAMPLE					ANSWER
88	21	79	78	16	282
QUESTIONS					
18	76	37	51	23	
73	70	27	50	35	
43	69	31	71	96	
63	79	48	12	13	
28	23	71	28	52	
31	64	48	48	45	

And there were these types of questions: 'A baseball bat and a ball cost $1.10 together, and the bat costs $1.00 more than the ball. How much does the ball cost?

Did you arrive at 10 cents? If so, I'm afraid you'd be wrong. (Find the solution in the references.)

The study revealed that a 1-degree Celsius increase in the room's temperature was associated with a nearly 2 per cent increase in the number of maths questions the women correctly answered, and a 1 per cent increase in their performance on the verbal task. The men, meanwhile, did better at cooler temperatures but their decrease in performance at warmer temperatures was nowhere near as great as women's gains.

Two per cent doesn't sound like much but there are some situations where it could make a big difference to an individual – for example in an exam, where it might mean the difference between an A and a B. And at group level, tiny increases aggregate to become significant differences. As the study's authors pointed out, a minimal tweak to the office thermostat can potentially lead to a clear win for group productivity and performance in mixed-sex workplaces.

Beyond the effect of these ambient tweaks, there is now a whole domain of behavioural sciences that focuses on nudging us towards making a particular decision. Under former British prime minister David Cameron, the Behavioural Insights team, popularly known as the Nudge Unit was established to apply behavioural science to government policy. The idea is that people can be gently moved towards making decisions in their own and society's best interests by altering their environment to steer them towards a particular choice. Since the default option around pensions became automatic participation rather than automatic opt-out, for example, 4 million more people in the UK now have some provision for retirement. Moving confectionery away from supermarket checkouts reduced the amount that parents bought for bored children as a result of pester power, cutting sugar intake significantly.

Changing the environment in which people make choices can steer them towards a particular decision without them necessarily being aware of it. Much of our decision-making is emotional and reactive, and we might not pick up on the factors that drive our behaviour. This kind of nudge engineering is infuriating to some people but is widely used to boost decision-making in the best interests of society as a whole: in public health and the tax system, for example.

Many of us would accept that we are all at the mercy of agents seeking to steer us in a particular direction – this is the basis of advertising, after all. The problem is, some of the ways that we alter our environment are more far-reaching and have more potentially unhelpful consequences than others.

Medicating Away from Joined-up Thinking?

In the UK, one in ten people over the age of sixty-five already takes eight different medications every week. There's increasing evidence that some of these medications are altering our behaviour on a vast scale. Statins, which are commonly prescribed to protect against heart attacks and strokes, have been linked to bursts of anger and volatile behaviour. Medications for asthma have been linked to hyperactivity and problems with attention; drugs used to treat Parkinson's disease with risky novelty-seeking and antidepressants with increased neuroticism. Even everyday drugs we think of as completely safe, such as paracetamol, can affect our brains in ways that might undermine pro-social behaviour and so reduce our capacity to think in terms of the group.

Recently, there has been a barrage of data that shows how commonly prescribed or even off-the-shelf pharmaceuticals can profoundly alter mood and behaviour. This is a massively understudied issue. All drugs must of course go through a series of clinical trials to ensure that firstly, they work and secondly, they don't produce toxic side effects. These trials are generally rigorous and highly trustworthy but they don't always investigate effects on the brain. Things like liver toxicity, kidney functioning, muscle function and blood protein in the urine are all routinely investigated, but effects on neural-circuit function or neurochemical transmitter levels are not. In essence, drugs are not routinely screened for effects either on the brain or on behaviour before their approval for use. This both baffles and concerns me.

Paracetamol is the go-to option for general pain relief for millions of us all over the world, but we are only now starting to discover some of its side effects. It reduces physical pain by decreasing neuronal activity in the insular cortex and dorsal anterior cingulate cortex, brain regions that play an important role in emotions, empathy and decision-making. Recent research demonstrated that paracetamol can also alleviate feelings of emotional pain such as unhappiness following a rejection, which led Dominik Mischkowski, assistant professor of psychology at Ohio University, to wonder whether it might also blunt compassion or diminish the ability to share other people's emotions. So in 2019, he and his colleagues recruited 114 students and split them into two groups. One group received a standard 1,000mg dose of paracetamol, while the other was given a placebo. Then he asked the participants to read about uplifting experiences that had happened to other people: 'Suzie had worked hard to get the job she has now. She can finally take care of her son and has almost saved enough to give him the gifts he wants for his birthday. Today in the office Suzie's boss told her that she had earned a raise for the great job she was doing.'

The results revealed that the single dose of paracetamol reduced people's ability to feel empathy and positive emotion by around 30 per cent. This has worrying implications, given that 6,300 tonnes are sold annually in the UK, which equates to seventy paracetamol tablets per person per year. The US buys 49,000 tonnes each year, equivalent to about 298 tablets per person.

It's not just painkillers that we might need to be concerned about. Commonly prescribed antidepressants, particularly selective serotonin reuptake inhibitors (SSRIs), are highly effective for millions of people in the treatment of depression and anxiety but they are also known to blunt emotions, and emotional numbing is a significant predictor of apathy. Emotional detachment might be a welcome relief for someone having to deal with trauma, but detachment from the world at large and other people in particular has negative long-term implications for individuals and the groups they interact with.

This is another area of interest to Molly Crockett and the team at Yale University. They conducted a series of experiments looking at how SSRIs impact our empathy and moral decision-making, not just in people who have been diagnosed with depression and prescribed the drug, but in anyone exposed to it.

Molly gave healthy volunteers either a single dose of the SSRI citalopram (also called Cipramil) or a placebo. She then asked them to consider the classic trolley dilemma. This famous thought experiment has been extensively used in studies of moral decision-making. The scenario goes like this: there is a runaway trolley heading down a track towards a group of five people. If it hits them, it's certain to kill them. You could pull a lever that would divert its course onto a different track, where there is just one person who would similarly be killed if they were hit. Participants in the study are asked to decide whether they would pull the lever, condemning that one person to certain death, in order to save the other five, or take no action. The experiment can be reframed in various ways to increase or decrease the amount of agency people feel: by, say, replacing the lever with another person, whose body would stop the trolley's progress if you pushed them into its path, but who would of course be killed.

This experiment has been run countless times and generally speaking, most people are much more prepared to pull the lever than they are to push a person into the course of the oncoming trolley. In either scenario, their action means that one person will be killed in order to save the lives of five others, but they are evaluated very differently. When the intervention is via a lever, the principle of utilitarianism generally pushes people to calculate that sacrificing one life is a worthwhile trade-off for saving five others. It's much harder to conclude this if you have to imagine putting your hands on a person and shoving them to their death.

This task is cognitively challenging. Our brains have to work hard to think through the various possibilities and scan them against our beliefs. The calculation raises people's stress levels. There is always a significant proportion who do nothing and so allow five people to be killed.

Crockett and her team discovered that a single dose of citalopram made participants less willing to harm one person in order to save many others, even in the first scenario involving the lever. My standpoint, in life and science, is firmly utilitarian. This inclines me to make decisions according to the idea that they should deliver the greatest amount of good for the greatest number of individuals, all of whom are presumed equal. Viewed through this lens, Molly's result rang alarm bells – the citalopram seems to diminish an individual's ability to engage with the complexity of the problem, to do the mental work of feeling their way around it. It increased their detachment and produced far more instances of the bystander effect.

Crockett's rather understated summing up was that, 'In the coming decade it will be important to systematically investigate these questions and debate their significance for morality'. Given that figures show 17 per cent of the UK's adult population was prescribed antidepressants in the year 2017–18, and that proportion has certainly gone up since the pandemic, the question of how these drugs are affecting our emotional intelligence and moral decision-making is crucial.

This is especially true when you consider that it's not only those people who've been prescribed these drugs who are affected by them. A significant amount of chemical residue passes into the sewage system and eventually leaches into the waterways. Anyone drinking water from the tap ends up ingesting it, with unknown effects on behaviour. A study conducted almost twenty years ago by the US Geological Survey found measurable amounts of one or more medications in 80 per cent of the water samples drawn from a network of 139 streams in 30 states. The drugs identified included antibiotics, antidepressants, blood thinners, heart medications, hormones and painkillers.

Scientists have been sounding the alarm for years about the negative effects on the reproductive biology of fish and other aquatic creatures caused by hormonal contraceptives and testosterone creams that end up in our rivers. Recently, the effects of pharmaceutical pollution on behaviour have begun to be investigated. Tomas Brodin, professor at the Swedish University of Agriculture, was interested in the effects of psychiatric drug pollution on aquatic life. He exposed a number of fish species including perch, roach and Atlantic salmon to oxazepam, an anti-anxiety medication that has been found to be polluting waterways throughout Europe. Compared with controls the medicated fish in his laboratory tanks moved around more, ate more and significantly reduced their social interactions. The normally fairly friendly species became less and less interested in interacting with others of their kind.

I asked Tomas if he thought the levels of psychiatric drug pollution in waterways might be affecting humans. 'There are absolutely sites in the world where, if you drink the water from the river, you will be medicated: that is, you will be consuming a dose that has a therapeutic effect on a human. These high concentrations have been measured fairly close to production sites or larger hospitals, though. As for drinking water from your tap, it probably contains several pharmaceuticals – but the concentrations are likely to be very low. That said, I almost never drink water from the tap except at home in remote northern Sweden, and I pay more attention to where the fish I eat were caught as well.'

Back in 2011, the Harvard Medical School wrote an open letter highlighting the topic of pharmaceutical pollution. 'Disturbing clues from aquatic life suggest now is the time for preventive action,' they concluded. Little action has been taken since then and in 2018, it was demonstrated that Atlantic salmon smolts exposed to anti-anxiety medications including Valium or Xanax are definitely being adversely affected. They migrate nearly twice as quickly as their unmedicated counterparts, but recklessly so, since the juvenile fish arrive at the sea in an underdeveloped state and before seasonal conditions are favourable. Drug pollution changes behaviour in ways we don't yet understand.

It's important to remember that all these medications decrease suffering for individuals and increase their ability to contribute their unique perspective to the group. Every person whose depression or anxiety is relieved by medication is freed up to think and behave in more rewarding and effective ways. I am not implying that there is any reason to panic or that anyone should stop taking their prescribed medication, which might have seriously dangerous consequences. But is there a risk that some of us might be losing a degree of our social intelligence, without even realising it? To my mind, this warrants investigation during the process of licensing drugs. It's all a question of balance. What are the downsides for individual and collective wellbeing and are we sufficiently aware of them, so that we can take steps to protect ourselves from them? As yet, I don't think we are.

Our Brains Online – the Good, the Bad and the Ugly

Digital information is ubiquitous in our world today; it's the water in which our minds swim. Social media, which trades in both information and emotion, is the most significant new component of our context for the emergence of either collective intelligence or collective stupidity.

Much has been written about whether the shift to online living is good or bad for us. I suspect that the answer is 'both'. The most exciting developments in joined-up thinking certainly could not happen without our digital network – one of humanity's greatest collective achievements – and few people who've come of age since its ubiquity can imagine life any other way. Millions of us have found jobs, love, friends, our tribe or our niche on the internet. But it can also be an alienating environment that encourages hostility and narrow-mindedness. Twitter and Instagram have been designed to reward fast rather than slow thinking, in which we're encouraged to react quickly, emotionally and irrationally.

I want to focus on the way our digital environment affects memory, decision-making and emotional contagion along with our ability to empathise, evaluate information, listen to others' points of view and form opinions. We'll have a look at how a shortened attention span, echo chambers and confirmation bias blend with emotional contagion to chip away at the skills we need to support pro-social behaviour, as well as critical thinking. The more we understand how our brains behave in this new environment, the smarter decisions we can make about how we live online.

Smartphones, Memory and Losing the Human Touch

Many of us have the nagging feeling that our brains are less agile now that we're so reliant on digital devices. We have outsourced our memories for facts and bus routes to Google, and for friends' phone numbers to our digital address book. We find our phones easier to interact with than our real-life friends and family and worry that we might be addicted, since our screen time just keeps going up. There's also the sense that our ability to focus on one thing for any length of time is diminishing under the pressure of constant scrolling and refreshing. And all this is before we get to the specifics of social media and what that's doing to our brains, or to our children's brains. We read that Steve Jobs did not allow his children to have iPads and wonder whether we should follow his lead.

It's hard to know how seriously to take these feelings, especially when our relationship with the technology is still in its infancy and when even the experts seem to disagree. In 2019, during a discussion hosted by the Royal College of Psychiatrists' International Congress, a delegate is reported to have asserted that 'excess screen time has reduced our attention span to eight seconds, one less than that of a goldfish'. Psychologist Dr Amy Orben of Cambridge University, who commented on the 2019 congress, was unconvinced. She called on Royal College of Psychiatrists delegates to show the evidence to support their claims and was trenchant in her pushback against the narrative that smartphones are addictive, saying, 'There is a widespread belief that [they] cause a dopamine kick and dopamine kicks lead to addiction. Well, anything I do that is pleasurable will give me a dopamine kick . . . I could be talking to my friends or eating a pizza.' She compared our relationship with our smartphone to our relationship with food. 'Sometimes I use social media to escape from a problem, and I know it's not good for me, but as with the mince pies at Christmas, I try to self-regulate.'

Dr Orben has a point – the addiction trope is often overused – but though she may be able to have a healthy relationship with both food and technology, many of us can't. Human beings are extremely susceptible to anything that hijacks the brain circuits that determine motivation and reward. It may take more or less repetition of the hit, depending on the substance and the particular genetic make-up of the person who's exposed to it, but for some of us, sweet foods such as mince pies will do it, as will smartphones.

When it comes to food the picture is complicated, with numerous factors converging, but it is clear that many of us struggle to control what we eat now that calorie-rich, heavily processed food is so readily available. The obesity epidemic is proof of that. Are technological developments making our brains flabby, just as modern food is making us obese? Quite possibly.

Memory is one area where we can see its effects. Memory is of vital importance to all intelligence. If we can't recall information, we can't use it. Even more fundamentally, we can't learn without being able to compare new information to old and update accordingly. We can't generate our sense of who we are without memory either, as anybody who has witnessed the devastating effect of dementia on a loved one will attest. Memory is crucial to all cognition and yet it's one of those functions that is both costly and undervalued until it's gone. Human beings have always looked for tools to make it easier to remember – rhymes and acronyms, marks on a wall, paper and pen, voice notes on our phone. We outsource it to other people, either explicitly via an assistant who keeps our calendar, or unconsciously via the transactive memory bank we instinctively form with our life partners through conversation, sharing the cognitive load in an effort to enhance brain capacity.

When transactive memory was first proposed, in 1985, the internet was only starting to become accessible to members of the public. Transactive memory was different from all the static forms of external memory storage, via address books and the like. It was the dynamic back-and-forth communication between people that was crucial to the way transactive memory allowed memories to be recalled, and to evolve. But now, of course, the internet is also a dynamic, interactive, shifting, stimulating beast that can engage in a transactive process in much the same way that people can. Most of us now rely on it more than we do on actual people.

Daniel Wegner was a professor of psychology at Harvard University and the person who developed the concept of transactive memory. His studies showed that by expecting to have constant access to instant information, by leaning on Google rather than relying on our own brain power or a trusted partner's, our personal rates of recall are reduced. We lose both the desire and the ability to store information in our own brains.

This might not be a problem – after all, why waste cognitive capacity on mundane memory when Google can do it for you – were it not for three things. Firstly, what happens when your phone battery dies? Secondly, as we've said, memory underpins many other vital cognitive skills that we still can't outsource and probably don't want to miss out on. And thirdly, if we give up on the transactive memory we form with other people, we're also deskilling ourselves for collaboration with them. Our capacity for collective intelligence goes down.

Although digital communication undoubtedly helps millions of us to keep in touch across the globe and played an important role in combating isolation during the coronavirus pandemic, its lure can be dangerously powerful. The mere presence of technology creates a so called 'virtual distance' where people are physically together but detached, absorbed by their own individual screen. Human-to-machine interactions have been shown to reduce intimacy and have a tendency to replace many human-to-human interactions. Our ever-growing reliance on technology seems to be undermining human collective intelligence in everyday life.

Last year, when I was living in Australia, I took a trip on public transport to go to a meeting in a different town. It was all unfamiliar and I'd forgotten to map my way from the bus stop to the office before I set off. My phone wasn't working so I asked a stranger, who was getting on the same bus, for directions. It turned out she was a palliative care nurse who had just finished her night shift. She had also just drunk a double-shot latte and was *very* awake. We started chatting about her job, about life and death. An hour later, after a mammoth, fascinating and strangely uplifting conversation about the process of loss, the frailty of life, the health culture in Australia, fate and biology, we disembarked to our respective destinations.

I certainly learned something from her – a refreshing view on how close contact with death can provide the necessary perspective to make the most of life, to stop sweating the small stuff and appreciate what you have. (I'm not sure she took anything nearly as profound from me!) This chance interaction, which I suspect affected me far more than any podcast I might have listened to could have done, would probably never have happened had my phone been working.

Travel through public spaces provides opportunities to bump into people outside our own bubble of experience. In theory, technology allows us to travel across the online world, to meet people with fresh perspectives and knowledge. In practice, as we know, the algorithms that build our online world are turning us into polarised tribes. Civil communication is breaking down. There are too many echo chambers where conspiracy theories breed. There's a constant stream of outrage and fear, spreading their emotional contagion across the network in seconds. These things are bad for our individual brains and function as barriers to collective intelligence.

But knowledge is power. The more we know about the risks and benefits of depending on digital, the more we can bring our full range of reflective decision-making to bear on how we interact with it. We can use timing settings to nudge us away from endless scrolling, remove apps from our phone, practise mindfulness without our phones by our sides. There are many intelligent ways to relate to the online world that bring us more of its upsides and fewer of its downsides. For the sake of our brain health, we have to use our smartphones cleverly.

Fear: Too Much Bad News Makes Us Stupid

To be alive in the twenty-first century is to live with a barrage of mostly bad news. Whatever happens to be going on in global events, whatever the relative happiness of our own little corner of the world, there is no escape from information and emotion now that we all inhabit a 24-hour news cycle delivered to our brains round the clock by our smartphones. From mass media to friends' social media posts, the default condition is to be bombarded constantly with urgent updates.

Part of the problem is the sheer volume of this tidal wave. Another aspect is its prevailing mood, which tends to the sensationalist, divisive and distressing. It has been asserted that Facebook and Instagram incentivise negative inflammatory content in order to capitalise on our appetites for the dark side of life. Former CIA analyst, diplomat and Facebook whistleblower Yaël Eisenstat has said that, 'As long as [social media] algorithms' goals are to keep us engaged, they will feed us the poison that plays to our worst instincts and human weaknesses.' A recent survey by the American Psychological Association found that although more than half of Americans say the news causes them stress, anxiety, fatigue or sleep loss, one in ten adults still checks the news every hour, while 20 per cent of Americans report 'constantly' monitoring their social media feeds.

We know that emotional contagion takes place via exposure to media of all kinds, from emojis to live performance, and that online platforms have simply magnified this phenomenon. We also need to be aware that negative moods are more contagious than positive. A small study conducted by Dr Per Block and Dr Stephanie Burnett Heyes at the universities of Oxford and Birmingham respectively looked at the way that moods were transmitted between adolescents on a week-long residential course. They asked participants to complete mood ratings and tallied their social interactions over the course of the week. What they discovered was that people were indeed likely to 'catch' their social contacts' mood, but that infectivity rate was substantially higher for bad moods than good ones. On the plus side, the person with the bad mood was likely to report an improvement in their mood from the diluting effect of spreading it across the network, which suggests a mechanism for emotional buffering. On the downside, for those who caught the bad mood there is a clear cost of offering this social support. (Worth remembering if we're stuck in a loop with a particularly toxic friend and wondering whether we're exaggerating the way it makes us feel.)

Offering solace to a friend or community member, even if it means that we have to take on some of the burden of their distress or fear, makes sense if human beings are only circulating in relatively small groups and have control over the amount of information and emotion that arrives at their doors. But our environment is different now. As Stephanie pointed out to me, 'social-media-facilitated networks are not just larger, they are also less close and supportive . . . [and this is very different to] . . . small groups where everyone is accepted and valued, and group members are working toward a common goal'. If we're plugged in all the time, exposed to messages that tell us the world is hostile, scary and unpredictable, naturally we find this alarming. And fear is terrible for our brains. It induces individual and collective stupidity.

Being exposed to a constant drip feed of anxiety-inducing news can profoundly affect our brains and behaviour. The problem comes from never getting to discharge those stress hormones. Overcoming stressful events can in fact make us stronger, enhancing our resilience and sense of self, but chronic low-level stress such as that brought on by doom-scrolling can be detrimental to both bodies and brains, as the respondents in the American Psychological Association survey reported.

Stress is a major trigger for persistent inflammation in the body, which can result in diabetes and heart disease. It can also overwhelm the brain. Normally, our brains are protected from circulating inflammatory proteins by a blood–brain barrier, but under repeated stress this barrier becomes leaky. The hippocampus, a key area involved in learning, memory and directing how we navigate the world, is particularly susceptible to their effects. We can get stuck in a rut of negative thinking, and this rumination decreases the potential for flexibility in thinking. We become less and less able to solve our problems with positivity and pragmatism, and more and more apathetic and depressed. This creates attentional bias to negative information, so that the world becomes an increasingly demoralising and stressful place to live in.

Chronic stress can also hand control to the brain's system of fast thinking, putting our so called 'hot and cold cognition' out of whack, which interferes with the balance between our rational thinking and emotions. It causes greater activation of brain areas such as the putamen, which is integral to motor control, speech, language and cognitive functions and thinking about ourselves. High cortisol levels are also associated with shrinking the brain, which hampers how we take on board new information; the crucial hippocampus is, again, particularly susceptible to this. Just to round off the devastation, elevated cortisol can also mess with our body clocks, impact our sleep and cause toxins to build up in the brain as our moods become increasingly despondent.

When we combine the way that digital media has been designed to keep us scrolling with the fact that negative emotions are more sticky than positive ones, plus the devastating physiological effects of constant low-level anxiety, it's not difficult to see how we can end up stuck in a loop of doom at 3a.m. every night.

This is bad for the individual and also for the group because as we've seen, fear and stress generally make people more insular, suspicious and untrusting. This very much depends on the individual person, but none of us is immune. Numerous studies have shown that fearful people struggle to empathise with others and plan for the future. Their bodies and brains are trapped in survival mode, which, simply put, means they are so focused on immediate threat that they've no chance of contributing to any project beyond that. This is why it has always been a tactic of war to demoralise and exhaust people.

Our digital technologies are still new and we are still grappling with how to use them. It's up to us to steer towards the positive forces they offer and away from their toxic pull. On an individual level, it makes sense to switch off from constant exposure whenever we can. We need to learn to consume digital information mindfully.

Collectively, we might want to push for legislation to set boundaries on what content can be distributed. The internet idealists stress that information must flow feely, but as this technology grows into its awkward adolescent mode there is a pragmatic argument for saying that some controls would be beneficial. The world's first social media law was implemented in April 2019, in response to the shooting of fifty-one people in a Christchurch mosque, footage of which was livestreamed. Australian attorney general, Christian Porter, said that Facebook and Twitter 'should not be playing footage of murder', given that commercial television stations were legally bound not to show it. The legislation now stipulates jail penalties for the executives of platforms that spread violent content.

Polarisation and Hostility: How We Destroyed Debate

Alongside fear and stress, digital life has also opened up a channel for aggression, which overlays and feeds the polarisation of viewpoints. Newsfeeds fill up with bullying, threatening, inflammatory and derogatory comments, as people hide behind their screens and become desensitised to interactions that were previously governed by social norms of respect and politeness. It seems to be ever easier to feel and express outrage online at all the stupid, absurd or horrendous things that *other* people do, say and believe.

Part of the responsibility obviously lies with us. Jamie Bartlett, author and director of the Centre for the Analysis of Social Media at the think tank Demos, specialises in online social movements and the impact of technology on society. He makes an interesting point about the part we all play in polarisation of opinion and retreat into echo chambers. 'One of the worst things about total personalised media is the conflation between "the media didn't cover that" and "I didn't see it in my feed". Everyone complains about the biased media, not realising it's usually their own curation bias that's the problem.' In other words, if we are to be capable of participating intelligently in groups of all kinds, then we need to actively work for balance in our information sources. In a world of polarised opinions and fake news, that means having the humility to be aware of our confirmation biases, and acknowledging that we're not immune to emotional contagion via inflammatory material. We need to seek out alternative views and be curious and sceptical, though not cynical.

Much of the responsibility lies with the social media companies. In 2021, a *Wall Street Journal* article reported on an internal Facebook presentation from 2018 that, they say, pointed to the way the company was encouraging growth in extremist groups' presence on their platform. 'Our algorithms exploit the human brain's attraction to divisiveness,' read one of the slides from the presentation, which warned that, 'if left unchecked', Facebook would feed users 'more and more divisive content in an effort to gain user attention and increase time on the platform'. Executives were alleged to have largely shelved the research but in a statement following the report, Facebook denied the research was definite, rejected suggestions they sought to polarize people and said they continued to work to take steps to address the issue.

All over the world there is a rising tide of intolerance for different views. This increasing tribalism doesn't all spring from the internet, and neither is it confined to online debate, but the technological capability of an infinite information network has definitely allowed it to spread. Now it seems to be the way we talk to each other. Or rather, shout. It's closing down our shared belief in mutual respect, freedom of speech and independent enquiry.

In March 2019 Jordan Peterson, professor of psychology at the University of Toronto, announced that he was joining Cambridge University on a two-month sabbatical with the Divinity School. Peterson's academic work, focusing on the psychology of religion and social conflict, is both prolific and heavily cited by fellow academics. He is also a highly controversial figure with a big platform. He's the author of bestselling self-help book *12 Rules for Life: An Antidote for Chaos*, has amassed more than a million followers on Twitter and has been vocal with his views on gender (including his refusal to use any pronoun other than he or she), as well as accusing climate movements of becoming too politicised.

After tweeting about his imminent sabbatical at Cambridge, a photograph came to light that had been taken after one of his public lectures. Peterson was standing next to an audience member who was wearing an overtly offensive anti-Islamic T-shirt. There was no suggestion that Peterson knew the person, but there had already been vocal concerns about his outspokenness with his views, and the photo sent the issue stratospheric. The public outcry was shortly followed by the fellowship offer from Cambridge University being withdrawn. This move was supported by 41 per cent of students, while 44 per cent supported a similar ban of Germaine Greer for her views on gender.

The issue didn't stop there. As cancel or call-out culture took hold in universities and online, the university put out an amended version of its mission statement around freedom of speech. It called for all points of view to be 'respected'. This only poured fuel on the flames. Numerous academics expressed concern that this would stifle the diversity of views and independent enquiry it was setting out to protect. Then Stephen Fry stepped in, remarking to the BBC that calls for 'respect' might have been well-intentioned but people could not 'demand' that their views would always be respected by others. Eventually, following vigorous debate, it was decided that the policy on free speech would support 'tolerance' of differing views rather than 'respect'.

In the ivory towers there may be vigorous but polite debate over the safeguarding of free speech. Online, the discussion can be much less well-mannered. But how effective is it to respond with outrage or anger to views we find abhorrent or know to be untrue? Does it serve any purpose?

Victoria Spring, doctoral student in psychology at Pennsylvania State University, is interested in social emotions – particularly empathy and anger – and how emotions and group dynamics interact to produce moral judgements. Victoria and her colleagues have highlighted the 'upsides' of moral outrage. It can create positive social consequences by catalysing collective action. As civil rights activist and feminist Audre Lorde said, outrage can be a useful force in social movements.

Molly Crockett endorses Lorde's perspective – up to a point. 'Focused with precision, [outrage] can become a powerful source of energy serving progress and change.' But Crockett has questioned whether the upsides of outrage still apply in an online social media context. It remains unclear whether online anger can catalyse collective action for social good, especially given evidence that outrage benefits the political right more than the left, men more than women and white people more than racial minorities. Perhaps online outrage actually reduces the effectiveness of collective action by limiting participation if it switches people off? Crockett argues that anger – a key component of outrage – has been shown to impair strategic decision-making by reducing the ability to consider long-term consequences and assess risks. It enhances distrust in others and increases the tendencies to place responsibility elsewhere and oversimplify complicated issues.

There's also evidence that social media lowers our threshold for expressing outrage, since its social cost is much smaller than it would be if we got angry in real life. So there really is more anger online; and that anger, like any emotion, can be transmitted far and wide. Molly argues that the public sphere risks getting so inundated with outraged noise that important issues are lost or ignored. Online outrage becomes performative, a lazy form of empathy uncoupled from altruistic action. Or it pushes us to block or cancel those we don't agree with, a furious disengagement that creates dangerous echo chambers of uniform opinion and conformist thinking.

Propaganda and Fake News: Weaponising Information

The emotional responses of fear and anger undermine our brains' slow reflective thinking system, and erode our ability to think calmly and critically. We end up at the mercy of our own emotions, which are constantly being nudged by contagion spilling over from elsewhere. Thinking for ourselves can be extremely difficult online. Participating in online debate or posting on social media makes us vulnerable not only to emotional contagion but to manipulation by ever more sophisticated propaganda machines.

Whether it's commercial companies wanting to sell to us or political parties wanting us to vote for them, being online leaves us exposed in multiple ways. Much of our online interaction now takes place in echo chambers, so the possibility for diverse points of view generating collective intelligence is diminished. The solidarity that can protect a group from manipulation is also eroded when data-harvesting techniques make it possible for a person or agency to target messages in highly personalised ways. Online we are always alone, even if we're participating in a huge group on a worldwide platform, because our data exposes our every thought to anybody who wants to exploit it.

It's hardly surprising that we're living in an age of conspiracy theories when we know that governments and corporations now have the tech to confect their message and then pitch it to us with absolute precision. This capacity to control information while remaining hidden is alarming and it's perfectly rational to see it as a threat. Unfortunately, that assessment can too easily tip into delusional thinking. The combination of propaganda with manipulation intensifies mutual suspicion and ups the stakes in the information war. Fake news proliferates. A dangerous situation is created in which trust in the mainstream is eroded and smaller tribes of dissent coalesce. Echo chambers allow delusional theories to emerge, amplify and spill over. These information dissidents will be increasingly dismissed by those outside the group, even if some of their beliefs carry some weight, and eventually an avalanche of tribal tendencies buries all consensus beneath it.

If digital technology is the ecosystem that allows emotional contagion and misinformation to coexist on a massive scale, the Covid pandemic was the accelerating event that sent conspiracy theories into hyperdrive. When our environment is filled with stressors and uncertainty, our judgement is impaired. It is difficult to analyse incoming information when our brains are being bombarded with unprecedented events. In addition to this confusion and sense of being overwhelmed, a state of enforced isolation exacerbates the brain's tendency to harmful rumination. If we are cut off from contact with other people who might offer a different perspective, it's all too easy to fall into the echo chambers of the internet, where there are few checks on our speculations.

Even as the world moves away from the peak of the pandemic, uncertainty seems to be the general tenor of the times we live in. Intersecting crises converge, meaning that our environment becomes more and more complex but our brains' fundamental need to make sense of the world never lets up. There are plenty of interested parties happy to rush in with content that promises to help us understand, and not all of it is reliable.

Compelling conspiracy theories are not just about the information they convey, even if they promise to provide us with all the answers. They're more emotional than that; more wrapped up with social pressure and conformity as well as the pleasures of being part of a tribe. The allure of a community made up of people who share similar beliefs is increasingly tempting during times of threat. Belief in conspiracy theories is a protective mechanism that promises to satisfy our brains' need for answers and allows us to be part of the elite group of 'those in the know'.

It's extremely difficult to challenge misinformation and conspiracy theory once it's taken hold of hearts as well as minds. Professor Sander van der Linden, director at Cambridge University's Social Decision-Making Lab, explained why in an article on website TheConversation.com. 'Conspiracy theorists see patterns everywhere – they're all about connecting the dots. Random events are reinterpreted as being caused by the conspiracy and woven into a broader, interconnected pattern.' Once the brain gravitates towards a theory, it is prone to stick to it, despite any contradictory information that might emerge. This confirmation bias combines with the drive to be right and to defend against attack by doubling down on the belief. The result is that challenging somebody in an attempt to debunk their opinion typically has the opposite result from the one we intended.

We've probably all witnessed the stand-off that occurs when someone highly confident in their decision or belief meets an opposing point of view. Debate, let alone compromise, becomes impossible. In 2020, Dr Max Rollwage from the Wellcome Centre for Human Neuroimaging, University College London published a study combining brain scanning with behavioural and neural modelling to investigate this behaviour. He found that being highly confident about your position leads to striking changes in the way the brain filters information. The higher your confidence, the more likely your brain is to literally filter out any electrical signals that contradict it. This selective neural gating of information results in our belief being backed up. Confirmation bias is not merely a label for a psychological process built on choices. It occurs at a neural level.

It can be helpful to remember that our own beliefs are also susceptible to this irrational spiral. Everyone's brain operates with confirmation bias, in which evidence against our position is selectively disregarded. Essentially, once we have built up a perception of the world, we ignore new information to the contrary. Changing the way we think is cognitively costly and might be socially costly too, if our opinions or beliefs are wrapped up in multiple frameworks within our minds and embedded in our social scenes, daily rituals and habits. People do change their minds, make different choices and alter their behaviours, but rarely as a result of a single aggressive challenge.

Kindness Is Our Best Defence

The picture of human decision-making that I've painted in this chapter is intentionally sombre but it doesn't have to be gloomy. In this instance, I believe that knowledge truly is power. When we know how biased and narrow-minded we all are – by design – how irrational and flawed our thinking processes are, how vulnerable to peer pressure and emotional contagion, we can foster resilience to these effects if we consciously decide to do so. We can protect ourselves against the toxic side of social media and begin to develop immunity to the barrage of anxiety-inducing news about our world. We can learn to enter into civil, rational discourse with even the most vehement holders of different points of view. We can support our brains to analyse incoming information even when it goes against our existing belief systems.

Our species' propensity for collective stupidity does seem to be facilitated by technology, which also creates new vulnerabilities. Perhaps rightly, we are nervous of some of these developments, concerned that they bring with them very definite risks to our mental health and the wellbeing of people we love, as well as our collective values and even the robustness of our democratic systems. Our brains are undoubtedly prone to bias and error and open to manipulation, but increasingly we are discovering strategies for working round these flaws by building our individual and group resilience, exercising our compassion and altruism, taking the time to notice this skill in others and emulate them and making a conscious choice to curate our lives so that positive influences dominate.

A multi-pronged approach will be needed, one that combines long-term strategies and short-term rescues. Professor van der Linden's work has shown that the strongest predictors of resistance to belief in misinformation are high levels of trust in scientists and decent numeracy skills in the general population. Both these things help provide a solid foundation for understanding evidence-based knowledge.

Critical thinking, by which I mean the ability to engage our brains' slower thinking system and ask ourselves questions about the source of the information, its purpose and plausibility, is a skill that can be taught. We learn it at home every time a parent shows us that we do not need to accept the playground bully's taunt that we are stupid or worthless. We learn it in school when we're taught to take information from a number of different sources when we're researching a project. We need to learn how to do it from a young age, so that when we need those skills we have them at hand. It is difficult to embed it in a mindset that is already hostile to authority figures and entrenched in its opinions.

We can also immunise ourselves against fake news and conspiracy theories, just as vaccines can be used to protect us against disease. Traditional vaccines work by injecting a weakened dose of a pathogen in order to trigger antibodies in the person's immune system. This confers resistance against future infection, since the immune system remembers the attacker and can more quickly mount an appropriate response. The same tactic can be adopted with information: by exposing people to weakened doses of misinformation, mental antibodies can be cultivated that help protect them against fake news in the future. Van der Linden describes how this method works in a variety of contexts, including increasing the use of mammography in breast-cancer screening, decreasing teenage smoking and raising vaccination take-up.

Supporting someone else (or ourselves) to let go of a belief is hard. Gentle erosion is always more likely to be successful than aggressive attempts to get somebody to 'see the light'. A simple correction will not be enough and patience is key. Once somebody has let go of the false narrative, there's a cognitive hole to fill, in order for the brain to re-establish internal coherence. A compelling counter-narrative is needed. Finally, research indicates that it is important to support the change of mind by emphasising the positive social implications for the group of the individual's hard cognitive work, whether it's the positive impact of getting vaccinated for the community, or the benefit for their family or friends. This appeal to emotional intelligence and social awareness can help the brain to feel satisfied with its new belief.

These strategies approach from the analytical side, but what about starting with the emotional skills we need to resist being sucked into collective stupidity? When we're stressed or frightened we can't think straight, so the first imperative is to develop strategies for feeling less fearful. Avoid doom-scrolling, take physical exercise, practise yoga, meditate, dig out the lavender sleep spray – whatever it takes to build up a body of habitual behaviours you can use to dampen the activity in your amygdala.

Once we've begun to dampen down our fear responses, we can work on building our capacity to notice and appreciate positive influences and emotions. Happiness is contagious, remember, but not as contagious as unhappiness. So it helps to choose (where possible) to spend time with joyful or peaceful people and on activities that replenish our energy rather than drain it. Our capacity to feel compassion is heightened by practice, which can include carrying out conscious acts of kindness and also noticing when other people are being compassionate. Choosing to read or watch stories of kindness works just as well, and in fact reading fiction has been linked with an increase in empathetic skills. Anything that increases your sense of optimism about what people are capable of and deepens your understanding of other people's struggles will boost your capacity to feel what cognitive scientists call 'moral elevation'.

This emotion increases the prefrontal cortex's control over more primitive emotional circuits responsible for the fight-or-flight response, so that you are able to deploy more executive decision-making rather than being susceptible to fear and threat. Moral elevation also raises oxytocin levels, lowers cortisol levels and increases neural plasticity, facilitating the integration of unexpected experiences into your understanding of the world. Collectively, this positive emotion of moral elevation helps to promote a 'pay-it-forward' mentality. Taking the time to perform and notice acts of kindness can literally help to spread compassion across society.

All these strategies are mutually reinforcing. If we are more optimistic and less fearful, our brains have more capacity to slow down our thinking and evaluate events in a less egocentric and more collectivist way. We become more available to the group and more able to contribute our ideas. Our pro-social behaviours support the kind of critical thinking that enables us to resist bias and error even in the face of uncertainty and fear.

There are grounds for optimism about our species' capacity

to resist the dark side. Literally, because it turns out that optimism is also an inbuilt trait that can be cultivated. Molly Crockett and her colleagues have been busy on this issue too, coming to the conclusion that the majority of us hold, to differing degrees, an optimistic learning bias.

Negative emotions may be stickier than positive ones and we may have a negativity bias that pushes us to remember and anticipate the bad over the good but, luckily, we also seem to have a cognitive mechanism for pushing against these downer tendencies. Crockett and team's study showed that people tend to update their beliefs in response to better-than-expected good news but neglect worse-than-expected bad news. Most people learn in this pro-optimistic way, with the exception of those who've been diagnosed with depression.

In a series of four studies, each testing between 83 and 285 participants, the scientists demonstrated that people's optimism emerges not out of self-interest but out of concern for others. By manipulating the degree of concern that participants had for an unknown person (by making that person identifiable or more likeable) the scientists were able to increase participants' vicarious optimism for them. It seems it's not just ourselves that we wish, hope and believe in the best for, but strangers too – so long as we can see their faces.

This capacity to identify with them – to see their humanity – is crucial. 'Optimism in learning is not restricted to oneself,' the team concluded. 'People exhibit an optimistic learning bias for identifiable strangers and, even more markedly, for friends . . . We see not only our own lives through rose-tinted glasses but also the lives of those we care about.' This tendency to filter out bad news and notice the good may be linked in with the neural basis of emotional contagion, helping to seed and transmit a positive outlook across society.

Most of us, most of the time, wish only the best for other people – so long as we have not succumbed to the conspiracy theories, emotional contagion, bias or apathy that could poison our minds. This trait is a corrective to our collective stupidity and provides a basis for the glue of selflessness that can help our groups, and ultimately our species, flourish. No room for complacency, then, but reason for optimism indeed.

Scenius and the Alchemy of the Group

In this chapter we've been looking at some ways in which we can relate to our environment in order to maximise the effectiveness and minimise the friction of our interactions with other people. We live in a heavily engineered environment that has a powerful effect on our choices, decisions and emotions. It can be helpful to have strategies in place to handle this.

But sometimes, groups feel more like alchemy than strategy. They can certainly be tricky – colleagues can be exhausting, leaders can be demanding – but when a group is working well, it can be glorious. Have you ever felt part of a group of people that just seem to be on each other's wavelength, into the same things, productive, harmonious, sparky and fun? There's a certain magic to some groups I've been lucky enough to participate in, where we bring out the best in each other. It's not always possible to say exactly how or why this happens. It can feel like a sort of collective charisma, where interactions feel rewarding, entertaining and easy, and a shared project takes on a life of its own.

Sometimes, a group comes together with the right mixture of skills and points of view, the perfect mix of complementary approaches, all in the right place at the right time. They often form through a shared creative impulse and frequently there's a dash of rebelliousness or contrarian thinking. Brian Eno, founding member of Roxy Music, music producer, visual artist and cultural theorist, coined the term 'scenius' to describe the way a scene can emerge around a certain creative or innovative group in a particular place and time. The term was a conscious riff on, and riposte to, the idea of 'the lone genius' model of achievement. 'I was an art student,' he said in 2009 at a festival in Sydney, 'and like all art students, I was encouraged to believe that there were a few great figures like Picasso and Kandinsky, Rembrandt and Giotto who appeared out of nowhere and produced artistic revolution.' Eno wasn't satisfied with this model of achievement and eventually came up with his formulation of 'an ecology of talent, out of which arose some wonderful work'.

Scenius, in Eno's conception, depends on lots of people, 'some of them artists, some of them collectors, some of them curators, thinkers, theorists, people who were fashionable and knew what the hip things were'. For Eno, scenius is both the group and the group mind. It's the intelligence of the whole and an ecology of ideas that gives rise to new thoughts and new work.

This idea speaks to the alchemy of collective intelligence, when a group's output far exceeds the sum of their individual parts. 'Scenes' are a focus for creativity and innovation but the output need not be artistic, as it was for the writers, artists and thinkers of the Bloomsbury Group in London in the early decades of the twentieth century, or the New York music scene of the 1970s. It could just as easily be the innovation hubs of Silicon Valley.

Kevin Kelly, co-founder of *Wired* magazine, picked up on Eno's conception and developed it. He suggested that for scenius to emerge, a group needed a shared language, an enthusiasm for both risk-taking and subtlety, mutual appreciation and a dash of friendly rivalry. New tools and discoveries should be freely circulated and all success should belong to the group. He suggested that although scenius could theoretically emerge anywhere that a group could gather – in the corner of an office or a corner of a city – it was most likely to be somewhere overlooked or marginal. Its core members were likely to have a shared passion that made them peculiar to those on the outside, and the group would need a buffer zone of tolerant outsiders who weren't too perturbed by maverick goings-on and would act as a shield against the attentions of, as he put it, 'do-gooders, bosses, the police or any other interfering authorities!' He believed that although every university or start-up aspires to be a hub for scenius, its emergence is serendipitous and cannot be engineered. If we're lucky enough to be part of it, we can be grateful and curious about where it will take us. Otherwise, the best we can do is leave it alone. 'When it pops up, don't crush it. When it starts rolling, don't formalise it.'

There is something rather beautiful, albeit idealistic, about the idea of a marginal space full of mavericks co-creating a new take on rock music or computer programming. Kelly seems to me to be right that exceptional collective achievement can feel almost magical and is probably at least partly serendipitous. And yet many of the factors he and Eno name are precisely those that we've identified as being behaviours we *can* cultivate and optimise. First, bring together a group of open-minded and engaged people from a wide variety of different backgrounds. Define a shared purpose. Create a culture of curiosity and mutual appreciation built on gratitude for support and effort, as well as sharing of resources, ideas and success. Add a dash of friendly competition, stir well and let the magic happen.

EXERCISE: Reflect on and Repeat the Positives in Life

For one full week, dedicate ten minutes at the end of each day to write a list of the interactions or moments that left you feeling positive. These could be conversations that made you feel joy or sparked intrigue; spaces in which you felt calm or happy. On the other side of the page, list instances where you felt negative emotions, be they frustration, irritation, disappointment or anger.

At the end of the week, reflect on your lists and use them as prompts for decisions about how you behave over the following week. Go back to those spaces that made you feel positive. Seek out similarly positive exchanges; arrange to meet up with the people who left you feeling good. As for the list of negatives, avoid those people, places and situations, or consider how you could structure them differently to increase your chance of experiencing them more positively.

CHAPTER NINE

Cathedral Thinking: Collective Intelligence at Scale

It took thousands of people and generations of human effort to build a medieval cathedral. The stonemasons who carved the gargoyles or designed the flying buttresses at Notre Dame in Paris laboured for years, knowing that they would never see the project finished. Not even close. They worked anyway. They passed on their skills to the next generation. They trusted that their contribution to the collective effort was important and would stand as a tiny part of a much bigger whole, for the benefit of millions of people not yet born.

You can say the same about any large, complex, long-term project that requires people to think about the needs and benefits not just of their contemporaries but of their descendants. Cathedral thinking is what built the James Webb telescope or the National Health Service or campaigned for civil rights. Humanity's success has been realised by many groups of people chipping away at a problem for years, decades if necessary, supported by their ability to think beyond themselves and to collaborate across different institutions, sectors and countries, and over time.

This is a romantic view, I know, and one that has to be balanced out by acknowledging that cathedrals – literal or meta-phorical – are not built solely of noble self-sacrifice and zeal for the common good. A lot of collective endeavour down the ages has been performed by slave or indentured labour, and demanded by an authoritarian leader. Participation has been the result of simple necessity – to earn a wage – or for personal reasons that serve an individual's needs and beliefs.

This doesn't mean, however, that we can't be inspired by the collective achievements of the past. It takes the same leap of imagination to build a cathedral as it does a genetic database like the one currently being compiled by a team at the Wellcome Sanger Institute, University of Cambridge. The researchers are sequencing the DNA of *every single life form in the UK*. It's a task so vast that it boggles my mind. The ultimate ambition is to sequence every life form on Earth. They are creating a knowledge database for the whole world, and for the scientists of the future to use. Julia Wilson, who is one of the architects of this inspiring project, pinpoints resilience, flexibility and faith as key attributes any group needs in order to tackle big projects that span institutions and generations. Such complex thought processes depend heavily on our species' perception of time, which is so sophisticated that it's essentially the ability to mentally time-travel into the past and future. We are the only animals, as far as we know, that have the cognitive capacity to remember the past in such extraordinary detail and imagine the future in such extraordinary colour.

'Mental time travel' may sound abstract but actually, we do it every day. Without it, the drive to satisfy our own short-term needs would continually trump our capacity to learn from the past and plan for our own futures – let alone our children's or their children's. So memory and imagination are key to all learning and planning, but especially when we're being ambitious with our joined-up thinking. Just as it's smart to think about what 'future you' needs when you're weighing up opening another bottle of wine, it's also smart to consider what our descendants need when we're evaluating whether to open another coal mine. Without the ability to reflect on the past and speculate about the future, we cannot work intelligently at scale. Our efforts at problem-solving will be trapped in short-term thinking.

Neuroscientists have recently started to contribute to the understanding of how this crucial aspect of our minds arises. The science of how our brains perceive time is very new. We can now observe the ways that three different circuits for its measurement are plugged into the rest of the brain. They connect to: the hippocampus, which is the seat of memory and vital to learning; the reward circuit, so that we can relate time to motivation and pleasure; the amygdala, which generates anxiety about the future; and the orbito-prefrontal cortex, which generates feelings of regret for the past and supports the executive functioning that draws conclusions and makes plans.

Thanks to this awesomely sophisticated network of trillions of synaptic connections between brain cells, human beings don't simply notice that time is passing, we have opinions and feelings about that fact. We experience consciousness, allowing us to reflect on past experiences, incorporate memories into our deepest emotions, form a personal awareness of our place in the world, make decisions to shape and direct our futures, and share our thoughts and hopes with others. Without memory and imagination there is no individual or collective identity, no 'we' thinking, no collective intelligence.

Future-focused Thinking: Building the Future We Need

Some jobs need more than a few years to finish. Some tasks are so vast that however big the group we assemble to tackle them, they can't be completed within our lifetimes. And of course, a truly successful project must be built to last. Planning for the future is one of the most fundamental tasks that successful individuals and groups must carry out. Whether it's securing our long-term family finances or working towards the success of the human genome project, many crucial projects require us to navigate an uncertain, multi-stage course between our starting point and a distant end point.

All long-term tasks rely on a complex range of cognitive skills that have time-awareness embedded in them. They depend on our capacity to pay extended attention to a problem, delay the gratification of our own present needs and pick ourselves up from setbacks in order to learn from them. These behaviours are pivotal for the success of our relationships as well as our projects, allowing us to sustain friendships for a lifetime, guide our children as they grow and be part of a community.

Many of humanity's greatest challenges require cathedral thinking. S. J. Beard is a senior researcher at the Centre for the Study of Existential Risk at the University of Cambridge. The centre is 'dedicated to the study and mitigation of risks that could lead to human extinction or civilisational collapse'. In an interview for BBC Radio 4, Dr Beard described one of the questions they're currently investigating. 'What will the habitat of the human species look like in 5,000 years? Will we be living on spaceships, or back in hunter-gather societies, or with some version of what we have now? Some of those scenarios are much more attractive than others, but all require us to use cathedral thinking, and to think about humanity as a whole.'

The problem is, collectively, we human beings are out of practice at paying attention, being patient, and thinking long-term and about humanity as a whole. The times we live in nudge us in the direction of short-term thinking, and our attention span is under attack. We live in the quick-fix, sugar-rush world of market-driven economics, with short-term election cycles and quarterly performance reviews. Fewer of us live near our extended families, more of us live alone and many of us are lonely. Having 'enough time' is all but a fantasy for many people in the developed world, despite our lives of comfort and privilege. So is having a community that we enjoy being part of.

In these circumstances, it is not straightforward to acquire and maintain the skills that support the deep work of collective intelligence, pooling of wisdom or tackling long-term and uncertain projects. It is easy to say we need cathedral thinking but harder to know how to reintegrate it into our mindsets. After all, cathedrals are the product of a very different age from our own. And yet, we must try.

The religious belief that inspired them is no longer the powerful motivating force it once was, but cognitive scientists are discovering more and more about the underlying mechanisms of motivation and morality in human beings. Could we exercise our capacity for collective intelligence by tapping into them?

The human brain is a machine for making meanings, after all, and many of us crave more meaningful activities than our work can provide. We are, to varying degrees, motivated by novelty and by developing mastery of a skill. Some of us long to lose ourselves in creative activity and transcend time and our self through the experience of being in the flow. Altruism and pro-social behaviours are part of our evolutionarily coded survival strategies, and provide pleasures and benefits for the individual at the same time as they boost collective intelligence. These are some of the many brain states that we can cultivate, as a way to push back against the short-term instant gratification that dominates our environment.

Making the cognitive effort to expand our circle of concern to take in other people, other tribes, other generations as yet unborn, even other species, is perhaps a form of morality that will equip us for our futures. As Jonathan Sacks says, 'It needs moral courage to say no to the things that are tempting in the present but ruinous in the long run: drugs, cheap plastic goods, cars for all, and the other ways in which we enjoy our present at the cost of our children's future. We need space in our lives to gather collective wisdom about the common good and to consider sacrifices now for the sake of benefit of generations to come.'

There is an intriguing question of faith here. We have to believe that we can do it. Days after the 2019 fire that engulfed Notre Dame cathedral, Greta Thunberg implored the European Parliament to adopt the thinking style and working practices that had underpinned its original construction in their approach to tackling the climate crisis. This is a task that perhaps calls for faith even more than it does for ingenuity. 'The faith to build the foundations without knowing how we will build the roof,' as she put it. We need the ability to do our bit and trust that others who come after us will do theirs. To accept that we cannot plan every stage of this task, but we can design our response to the challenge with flexibility at its core.

Flexibility leads to resilience. The capacity to adapt to a changing environment can mean the difference between recovery and disaster, success and failure. The work we produce must be capable of being repurposed rather than scrapped. We must stop designing inbuilt obsolescence. We must work in ways that can incorporate learning as we go. As Dr Beard puts it, 'Future-proofing is not about knowing what the future's going to look like, it's about making systems that are flexible and can adapt to changing demands that we can't envisage yet. That's how you build in resilience.'

This secular but faith-based long-term thinking is the opposite of the way that most big projects are currently run by organisations, corporations and states. Managerialism blocks out every stage of a task, its schedule and budget. What we need now is very different. We need a philosophy of exploratory work that values collaboration and trust. We need ambition for our species, not for ourselves. We need acknowledgement that cathedral thinking is cognitively hard but both necessary and exciting, and above all we need the willingness to try.

Julia Wilson of the Darwin Tree of Life gene sequencing project said, on BBC Radio 4, 'We deliver cathedral science. We start early when the tech isn't ready yet, but we know the tasks need to be done.' This sums up the joined-up thinking we really need. A willingness to start before we're ready and learn as we go, to work in the direction of a destination that will benefit huge numbers of people all over the world and in the future. It embraces risk but understands that resilience is the goal. It is a long way from Silicon Valley's creed of 'fail faster, fail better', which can be glib at best and irresponsible and unethical at worst.

Cathedral thinking in the twenty-first century is a leap of faith, not in any god but in ourselves. Can we be inspired by the efforts of our ancestors, to become better ancestors ourselves? Can we follow the example of traditional Native American decision-making and incorporate its concept of 'seventh-generation stewardship' into our planning? Can we be ambitious, not for our personal advancement but for our species?

In his 2021 book *The Good Ancestor,* philosopher and author Roman Krznaric rallied us to think this big by speaking about justice. 'Our political systems disenfranchise future generations in the same way that slaves and women were disenfranchised in the past. Future generations are granted no political rights or representation. Their interests have no influence at the ballot box or in the marketplace. This leaves them vulnerable to multiple long-term threats, from rising sea levels and AI-controlled lethal autonomous weapons to the next pandemic that lies on the horizon, whether naturally occurring or genetically engineered.'

S. J. Beard and their colleagues, including Lord Rees, the Astronomer Royal, recognised this fundamental injustice when they and a group of undergraduate students from Cambridge University pushed the government to establish the All-Party Parliamentary Group for Future Generations in 2018. Lord Rees said, 'Future generations are the ultimate unrepresented constituency. Many of today's children will still be alive in the twenty-second century. When we realise how much we owe to the heritage left by our forebears, it is shameful how short-term most decision-making is. This group seeks to focus on the longer term, so that future generations aren't left to cope with a depleted and more dangerous world.'

It inspires me to see so many people working together, across generations and skillsets and all over the world, to cultivate cathedral thinking and embed it in projects that will support collective flourishing for our descendants. Roman Krznaric refers to the people who underpin such future-focused projects as 'Time Rebels'. Japan's Future Design movement is an example of the innovative and practical work that happens when time rebels adopt cathedral thinking.

Their aim is to overcome short-term and narrow-minded political thinking in town planning. The first step is to invite a representative and diverse group of residents to help with urban design, reaping the benefit of collective wisdom by forming citizens' assemblies. These groups are given a range of relevant information by experts with different views and then asked to evaluate it and devise strategies.

What makes the Future Design citizens assemblies different from those that have been successfully used to tackle big tasks, such as abortion reform in Ireland or city council budgets in Colombia, is the emphasis on thinking of and for future generations. Half of the participants are asked to imagine themselves as residents from the year 2060. They are designated 'time lords' and given robes to wear to help them embody their role as representatives of future generations. The time lords routinely advocate for much more transformative city plans than their present-day-focused peers, in every area from healthcare investments to action on the climate crisis. 'What if Future Design was adopted by towns and cities worldwide to revitalize democratic decision-making and extend their vision far beyond the now?' asks Krznaric.

What if, indeed? This is a gloriously imaginative and yet pragmatic vision of collective intelligence, harnessing diversity of input and different cognitive strengths and putting them in the service, via truly collective democratic systems, of collective success over many generations. Future Design is one of many projects that make me believe we can be the time rebels that our descendants so desperately need us to be, in order to create the cathedrals of the future.

Lessons From Our Ancestors

Cathedral thinking and future-focused thinking are inspiring ways to frame how we can expand our groups even more, and think even bigger. Another way is by turning to the past and considering what we've learned from our own ancestors.

How is knowledge passed from one generation to another? Aboriginal Australians or Amazonian tribes such as the Yanomami people embody a legacy of skills and knowledge that equip them for life in their particular environment. Some of this learning obviously comes about through imitation, and via language in the form of explicit lessons. 'Don't eat these mushrooms, they'll make you sick. Do go out in springtime to look for these berries.' Then there's transmission of information, lessons and memories via storytelling, which embeds knowledge within culture. Aboriginal Australian societies have made this central to their collective thinking. They imagine the timescale available to those of us alive today to be infinitely vast, with knowledge passed down the generations through dreams and stories.

There is also a biological process by which knowledge can be passed across a group that spans generations, imprinting in our bodies and brains. Humanity has been taught, through evolution, about the threats our ancestors faced and the gains in knowledge they made, so that we might benefit from their experience. Incredible recent discoveries in the field of epigenetics are pointing at how lived experience can ripple down the generations so that the effects of collective intelligence and collective stupidity alike reach through time. The children and grandchildren of Holocaust survivors, for example, carry the biomarkers of their ancestors' pain and resilience.

Could understanding the biological processes that underpin how experiences are stored – and erased – in the brain and body help shape our behaviour now? Perhaps, if we know that what we think and do today will impact on generations to come, we can be mindful of passing on improved capacities for collective intelligence to future generations.

At its heart, this means embedding positive social behaviour rather than antisocial behaviour in our groups. If a group of people goes to war against another group, or decides to enslave them, that profoundly antisocial behaviour leads to trauma, both for the individuals directly affected and, as we will see, their descendants. Trauma is staggeringly bad for intelligence, both individual and collective. Frightened, demoralised, stressed, angry people don't make good decisions and can't tackle problems on any timescale beyond the immediate need to survive. When our timescale is reduced in this way, our cognitive horizon is severely limited. It's damaging for perpetrators, too. For as long as the focus is on aggression, all the collective brain power of the group is going on destruction rather than collaboration and creativity. Everyone's outcomes are diminished.

Trauma, when it's experienced by a whole class of people such as Jews or those descended from enslaved Africans, is the outcome of being on the receiving end of collective stupidity. Healing from it is a form of collective intelligence. Such healing draws on all the behaviours that underpin any successful group's collective endeavour. What can we learn from groups of people who have recovered from trauma with optimism, empathy and compassion? How do people move on from painful memories? How does forgiveness emerge? Is there anything we can do to ensure that it's resilience and wisdom rather than fear and stress that we contribute to our groups, and pass on down the generations?

To answer these questions, we'll take a look at the neurobiology of trauma and resilience in the individual brain before turning to epigenetics to examine how their biomarkers pass from one generation to the next. Neuroscience is now developing techniques that allow us to heal through erasing traumatic memories in the brain so they can't be recalled by the individual or passed on to their descendants. It's also illuminating the way that healing from the past in order to build a better future is one of the pinnacles of humanity's collective intelligence.

Embodied Experience: Turning Trauma Into Resilience

Over the past decade, neuroscientists have been able to contribute to the understanding that psychologists have been accruing for more than a hundred years. The headline remains the same: traumatic experiences, especially in early life, can be devastating for the development of a person's brain, their cognitive skills, wellbeing and life chances. That said, they don't have to be. There is no direct cause and effect between a traumatic experience and a negative outcome such as drug addiction, self-harm or any other behaviour. The question of why some people seem to escape the worst effects of distress and others do not has also been studied by neuroscientists looking for the brain basis of resilience.

The ACE study (into the impact of Adverse Childhood Experiences) was the first of its kind when it was set up by Dr Vincent Felitti and Dr Robert Anda between 1995 and 1997. They recruited more than 17,000 volunteer patients from the CDC-Kaiser Permanente Health Program in San Diego, where Dr Felitti worked. The researchers asked participants about their childhood experience of ten different types of adverse events, including emotional neglect, physical and sexual abuse, drug misuse and violence in the household. The study, which is ongoing, is following people throughout their lives to record their outcomes and has generated enough robust data for dozens of papers and presentations at conferences.

The results are shocking. For a start, childhood adversity is much more common than had been realised. Secondly, for many people, negative early-years experiences aren't something that they can simply 'get over'. Almost two-thirds of people in the study reported having experienced at least one ACE category before the age of seventeen. More than one in five people reported three or more ACE categories. Were they a particularly vulnerable group of society, you might ask? Seemingly not – study participants were mainly white, middle-class, college-educated Americans from Southern California who had access to healthcare. Similar findings have been observed across the world and across demographics.

Trauma causes physiological changes – alterations in brain development, the immune system, hormonal systems and even the way our DNA is read and transcribed. This impairs our ability to heal effectively, raising the risk of cancer and heart disease. It also affects brain anatomy and function, during the early years particularly, but throughout life. Decision-making systems, emotional development and propensity for pro-social behaviours are all negatively impacted.

I spoke with Anne-Laura van Harmelen, professor of brain, safety and resilience at Leiden University, about what this means for people affected. Early trauma 'places the child at between three and eleven times higher risk for disorders such as ADHD, aggression, depression and anxiety, drug and alcohol misuse and both attempted and completed suicide,' she told me. Anne-Laura pinpointed 'high levels of stress hormones and immune markers (inflammation)' as important culprits and said that although the effect is most profound 'on the developing brain . . . they can persist throughout life, even when the stressor was experienced for a limited period of time'.

Trauma also prompts early synaptic pruning, which typically occurs in adolescence but can happen up to a decade earlier in children with high ACE scores. This is the process that reconfigures the brain by bringing certain regions into play through speeding up connections between them. Impulsivity and risk-taking are temporary behavioural outcomes of this process, and it can lead to poor decisions and high exposure to risky situations, all of which is likely to be more concerning in an eight-year-old than an eighteen-year-old.

The effects don't end there. Increasing vulnerability to psychiatric disorders such as depression and anxiety might be linked to trauma blocking the expression of BDNF (brain-derived neurotrophic factor) – a chemical that supports neuro-genesis (new brain cell growth). A flourishing brain relies on BDNF but a number of studies have demonstrated that abuse or separation from the primary caregiver quashes its expression.

An individual does not even need to have experienced traumatic events directly for their brains to be affected by them. People who witnessed the terrorist attacks on the Twin Towers in New York in 2001, at a distance of up to about 1.5 miles from the site, continued to have neuronal scarring on their brains years later. Exposure to the terrifying events and emotional contagion in the aftermath, as panic took over, was enough to cause neurological damage. The collective fear on that day rippled through a huge group of people and, as we might expect, it left a lot of damage in its wake.

Trauma can be incredibly bad for us, basically. Quite apart from the devastating impacts on an individual's life, it also has negative consequences for groups. A person who is experiencing a lot of pain and anger is not capable of deploying their intelligence for their own maximum benefit, let alone for their family or community's. Their behaviour is likely to tend to the antisocial rather than the positive social. And there's increasing evidence that this legacy doesn't die with them.

The Legacy of Trauma and Resilience –
How Knowledge Is Passed On

In the aftermath of the Holocaust, many survivors of the concentration camps suffered from post-traumatic stress disorder, and increasingly it became apparent that their children were also severely affected. As one psychiatrist wrote in 1966, 'It would almost be easier to believe that they, rather than their parents, had suffered the corrupting, searing hell.'

At the time psychiatry was dominated by psychological analysis and it was presumed that children's symptoms were the 'result of having traumatised parents who may have been symptomatic, neglectful, or otherwise impaired in parenting'. Over the past couple of decades, research has suggested an additional factor: trauma can not only be passed biologically through a group of contemporaries, but also through a group over time, leaving its mark on descendants' brains.

This work has been carried out in a new field known as epigenetics, and has demonstrated that traumatic memories can be stored in the conformation of our DNA, leaving chemical residues that act like a volume dial on a particular gene, either dimming its expression or ramping it up.

This dial-change can be passed on with the DNA code itself. The product of lived experience in one generation doesn't die with that individual but goes on to shape the physiology of the next generation, impacting on their brains' development and their behaviours. A positive environment can leave positive epigenetic marks and contribute to a pool of collective wisdom, while a negative environment can leave genetic scars that alter the epigenome and limit capacities for collective intelligence. (It's perhaps worth emphasising that this mechanism is not triggered by occasional lapses in otherwise 'good-enough' parenting but by the sort of severe stress uncovered by the ACE study, or being caught up in warfare or other disasters.)

Much of the early epigenetic work was conducted in model organisms including mice and even worms (though as we will see, the findings appear to translate to our species). My favourite pivotal study is one that left the neuroscience community reeling when it was published in the journal *Nature Neuroscience,* in 2014. Carried out by Professor Kerry Ressler at Emory University in Atlanta, Georgia, the study's findings neatly dissect the mechanism by which an individual's behaviours are affected by ancestral experience, creating a pool of collective wisdom over generations. The results are so groundbreaking that the paper has been cited over 1,200 times since publication.

The study was carried out on mice and made use of their weakness for cherries. Typically, when a waft of sweet cherry scent reaches a mouse's nose, a signal is sent to the nucleus accumbens, causing this pleasure zone to light up and motivate the mouse to scurry around in search of the sweet treat. Rather meanly, the scientists took a group of mice and exposed them first to the smell and then immediately to a mild electric shock. The mice quickly learned to freeze in anticipation every time they smelled cherries. Then the researchers let the mice be: no electric shocks and no sweet smells. The mice had pups and their pups were similarly left to lead happy lives without electric shocks, though with no access to cherries. The pups grew up and had offspring of their own.

At this point, the scientists took up the experiment again. They wanted to know whether the acquired association of the smell of cherries with a painful shock could possibly have been transmitted to the third generation. It had. The grandpups were highly fearful of the smell of cherries.

How had this happened? The team discovered that the DNA

in the grandfather's sperm had changed shape, leaving a blueprint of the experience entwined in the conformation of his DNA. This altered the way the neuronal circuit was laid down in his pups and their pups, rerouting some nerve cells from the nose away from the pleasure and reward circuits and connecting them to the amygdala, which is involved in fear. In addition, the gene for the olfactory receptor activated by that particular smell had been demethylated (chemically tagged) in the sperm, so that the circuits for detecting it were enhanced. Through a combination of these changes, the traumatic memories cascaded across generations to ensure that the pups would acquire the hard-won wisdom that cherries might smell delicious but were bad news.

The study's authors wanted to rule out the possibility that learning by imitation might have played a part. So they took some of the mice's descendants and fostered them out, away from siblings, parents and grandparents. They also took the sperm from the original traumatised mice, used IVF to conceive more pups and raised them away from their biological fathers. The pups that had been fostered out and those that had been conceived via IVF *still* had increased sensitivity to the smell of cherries, and different neural circuitry for its perception. Just to clinch things, pups of mice that had not experienced the traumatic linking of cherries with electric shocks did not show these changes even if they were fostered out to parents who had.

The team had succeeded in pinpointing how learned behaviour could lead to neuroanatomical changes across the generations, massively speeding up evolutionary adaptation. This is the mechanism by which the lessons learned by our ancestors might have been passed down to us. Their wisdom allows us to avoid the unpleasant experiences through which they learned the hard way.

The most exciting thing of all, perhaps, occurred when the researchers set out to investigate whether this effect could be reversed so that the mice could heal and other descendants be spared this biological trauma. They took the grandparent mice and re-exposed them to the cherry smell, this time without any accompanying shocks. After a certain amount of repetition of the pain-free experience, the mice stopped being afraid and went back to being excited by the smell of cherries. Anatomically, their neural circuits reverted to their original format. Crucially, the traumatic memory was also erased in the behaviour and brain structure of a new generation of descendants born after the grandparents had been reprogrammed to feel their original delight in cherries.

Studies on Holocaust survivors and their children carried out by Rachel Yehuda, professor of psychiatry at Mount Sinai Medical School in Israel in 2020, revealed that the same mechanism of epigenetic change can cause trauma to be passed on in humans. Her first study showed that participants carried epigenetic changes to a gene linked to levels of cortisol, which is involved in the stress response. This was one of the first proofs that memory of experience can be a biological as well as a psychological legacy. In 2021, Yehuda and her team went back to carry out more work and found gene-expression changes in genes linked to immune-system function. The suggestion is that these changes weaken the barrier of white blood cells, which allows the immune system to get improperly involved in the central nervous system. This interference has been linked to conditions including depression, anxiety, psychosis and autism.

It's important to emphasise that we do not yet know how long the biological markers of traumatic memories can last in the human brain or the number of generations they could be passed along. Preliminary studies, admittedly in worms, show traumatic memories epigenetically cascading down at least fourteen generations. How such mechanisms unfold in the human nervous system is the subject of a lot of current research.

Creating Growth Out of Pain

What about the biology of resilience and recovery? If people can pass on cognitive vulnerabilities to their children, can they also pass on the brain's coping mechanisms for managing them? Why is it that some people can recover from even the most horrific life experiences, and others can't? And given that Kerry's studies on mice showed that damage could be reversed with interventions, can we do the same with people?

Karin Roelofs, professor of experimental psychopathology at Radboud University, has conducted some pivotal work in this area. She discovered that some people are biologically primed to be particularly sensitive to distress. Her work, which was published in 2021, was based on studies of 210 police recruits who were about to embark on basic training in which they would be exposed to multiple potentially traumatic events. She conducted MRI scans and a battery of cognitive tasks at the beginning of the programme and repeated them after sixteen months. She found that the more trauma a recruit had been exposed to, the higher the level of their post-traumatic stress symptoms and the more active their amygdala, but some people's amygdalas had increased in activity way more than others'. So what was going on in the brains of those who were really struggling?

It turned out that recruits who had lower levels of activity in the frontal cortex before they started training were more likely to report stress. This makes sense because the frontal cortex supports us to cope with life stressors by exerting so-called 'top-down' control over the emotional systems, and especially the amygdala.

If this was key to the problem, could it be alleviated? The researchers set out to see whether it was possible to exercise the frontal cortex. Volunteers had this region electrically stimulated, to increase cross-talk between the frontal cortex and other regions in the brain network involved in emotion regulation. Simultaneously they carried out the same task given to the police recruits in which they had to react to emotional faces they saw on a screen, controlling their automatic tendency to avoid angry faces and approach smiling ones.

The team found that the brain stimulation helped to increase the frontal cortex's control over the motor cortex and amygdala, dampening down their activity levels. This work might be the basis of a new way to instil greater resilience in people who are routinely exposed to trauma, such as emergency-service workers, doctors and relief workers in disaster zones. With more resilience come less distress, more adaptability to a fast-changing environment and more access to slower rational thinking rather than being stuck in survival-mode thinking.

Karin also found that those recruits who reported significant symptoms of stress had a smaller than average dentate gyrus region in their hippocampus, which is involved in learning and memory. The dentate gyrus is particularly significant because it is one of the very few regions of the brain that contain stem cells, which are capable of generating new neurons. These new cells, once integrated within existing circuits, can help to create new memories to overlay traumatic ones. A smaller volume of this region seems to make people more vulnerable to getting stuck in a stressful loop of reliving painful memories. A bigger dentate gyrus, presumably with more stem cells and a higher turnover of neurons, seems to mean it's easier for the brain to create fresh memories.

Karin proposed a simple intervention to improve the resilience of people with smaller dentate gyrus brain regions who are likely to be exposed to high-intensity stress in their lives or work. Her advice is to take reasonably strenuous physical exercise, regularly. Studies have shown that exercise plumps up the dentate gyrus and increases neurogenesis. Just twelve weeks of running three times a week has measurable benefits for both.

All this pioneering work is shining a new light on a familiar truth: that distress – even severe distress – does not have to lead to tragedy. It can in fact generate growth. There are measures that people take, consciously and unconsciously, that can have a protective and recuperative effect. My friend Emily would score three on the ACE test, putting her at high risk for negative outcomes, but although she is prone to binge-drinking she's also building more and more protective behaviours into her lifestyle. She has moved to the aptly named Sunshine Coast in Australia and she exercises, meditates and eats a whole-food diet.

The healing potential of diet, and specifically its influence on our gut microbiomes, is particularly exciting. As more studies are carried out on the gut–brain axis, we're learning about the links between digestion, mood and cognition. Our intelligence is collective even at the level of our own individual organism, since it's effectively a whole-body function.

Trauma biology is one of the latest fields to begin to investigate the gut–brain axis for new therapies. The focus is on the trillions of microbes, including bacteria, fungi, protozoa and viruses, that make up the microbiome that lives on, and contributes to, the body. Most of these bugs live in our guts, particularly in the large intestine, where they help to digest our food, regulate the immune system and protect against bacteria that cause disease.

It's now being suggested that having a healthy microbiome can protect us against some of the effects of adversity. In 2020, Bridget Callaghan, director of the Brain and Body Laboratory at UCLA, published a groundbreaking study looking at a group of 344 children, many of whom had been separated from their parents and either been adopted or entered a care home. These children reported higher levels of stomach aches, constipation, vomiting and nausea than was typical. The team analysed gut bacteria in a subset of those children and found that their symptoms were linked to their distinctly different gut microbiomes, which were much less varied than usual. The team then tallied this with brain profile changes. Those children who had the most altered gut microbiomes also showed unusual brain activity in response to emotional faces. They reported significantly more and longer-lasting anxiety than the control group.

'It is too early to say anything conclusive,' co-author Professor Nim Tottenham from Columbia University cautioned in the research release, 'but our study indicates that adversity-associated changes in the gut microbiome are related to brain function, including differences in the regions of the brain associated with emotional processing.' Bridget suggests that making changes to our diets to boost our intake of fermented foods and probiotics could help to heal the damage to the central nervous system and digestive system that's caused by significant prolonged stress. This is especially important in children, where the impacts can be most severe.

Beyond the proven impact on our emotional intelligence of eating a good wholefood diet with lots of probiotics, and taking regular strenuous exercise, there are more radical interventions under investigation. As we start to understand how memories are laid down in the brain, we are also learning how to tweak or even eradicate them. Just as the mice in the cherries experiment could first be made to learn that cherries were dangerous and then to unlearn it, human beings' stock of information about the world can be altered or erased. Fear-extinction, as this process is known, is a fascinating and rather sci-fi concept. Fear certainly makes us less intelligent in the moment and now that we know it can be passed on to our children there's a clear argument for intervening to dampen it down. But do we really want to eradicate it?

All emotions, including fear, are in some sense messengers trying to get our attention. They contain lessons we might need to learn. A memory that makes us fearful or a sense of dread we can't quite put our finger on might be alerting us to something crucial, either in our immediate context or in the form of ancestral wisdom. Fear, like intelligence itself, is a survival mechanism. That said, it is also crucial to be able to inhibit fear when it is no longer relevant. The brain practises fear-extinction all the time in order to override an old memory with a new evaluation. Fear-extinction is a crucial part of our ability to learn and to respond to new situations. The brain is constantly undergoing a dynamic shuffle, forming and erasing memories, sculpting and re-sculpting our identities in the present and shaping the story of our futures.

Professor Tim Bredy and his colleagues at the Queensland Brain Institute published some mind-blowing work in 2020 that established how the brain manages this process. They wanted to investigate the brain's flexibility in memory forma- tion: how quickly and efficiently it can override an old memory in new environments. What they discovered was how memo- ries can basically edit genes, in real time. The researchers demonstrated, in mice at least, that DNA in brain cells can flip its shape. In essence, the more acrobatic the DNA, the quicker a mouse can learn from a new memory being integrated into neural circuits, and the quicker they can eradicate that memory as new information comes in to indicate that the environment is safe again.

Further studies are being conducted to discover how these results might translate to humans, and hopefully to clinical treat- ment for individuals. This holds out the prospect of a new therapy for boosting our ability to learn, rather than freeze, under condi- tions of adversity. Even more exciting is the possibility, theoretical for now, that these studies could offer some relief to whole populations afflicted with trauma. As we learn more about how long the ripples of damage caused by antisocial behaviour can last, we are also uncovering ways in which we might be able to improve people's health outcomes and free up a vast amount of cognitive capacity.

Fear-extinction is not the only mechanism being investigated for its impact on our cognitive capacity. Joel Pearson, the intuition researcher from the University of New South Wales, has been collaborating with Dr Selen Atasoy of the University of Barcelona to examine the effect of psychedelics on resilience and learning. Their work builds on pioneering studies by Professor David Nutt of Imperial College, London. What Atasoy and Pearson have found was that LSD could have a whole-brain effect on all aspects of cognitive capacity. They used a new method of brain analysis called 'connectome harmonics', which looks at neural activity as a combination of waves of particular frequencies.

They discovered that when somebody had taken LSD, regions of the brain became connected to other areas that they don't usually work with. The scientists termed this 'repertoire expansion', and noted that the effect seemed to be structured rather than random. The total amount of activity across all brain regions also increased. LSD appears to pave the way for enhanced improvisation in brain activity, as dynamics become faster and more complex, and innovative arrangements emerge. Atasoy uses the analogy of jazz and suggests that psychedelics push our brains in the direction of experimentation and improvisation, as if your brain had taken its cue from John Coltrane or Charlie Parker. Creativity and innovation are boosted, along with positive mood, memory function, adaptability and pace of learning. Though this sounds positive, it's important to note that all the scientists involved in these trials stress that psychedelics are not risk-free. In some people they can trigger first-time psychosis and they can make existing psychosis worse, so they should only be administered under medical supervision.

Might this emerging field eventually yield us the ability to intervene at scale to boost a group's cognitive capabilities? (Perhaps we're already there, and this research underlines that those tech workers in Silicon Valley who have taken to microdosing psychedelics are on to something.) Could we use psychedelics not only in the treatment of psychiatric disorders, as Professor Nutt's research has already suggested, but also to alter epigenetic memories passed on from our ancestors? To recover from emotional damage, boost emotional intelligence, increase our ability to adapt to our environment, learn more, create more and support each other more effectively?

There is emerging evidence that psychedelics could be used in this way, at least in mice, offering a potential tool to cut off the inheritance of painful learning and reduced ability. We will need more research to answer these questions as they apply to humans. What we know already is that epigenetics offers a new way to understand the influence of our environment on behaviour. This has huge and exciting implications for the rate at which we can learn. It means that those of us who have had our cognitive capacities limited, through no fault of our own, can hope for relief and recovery.

Above all, it gives us grounds for optimism that human beings can adapt, quickly, to our rapidly changing world. It used to be thought that our genes were the ultimate drivers of our behaviour. This was increasingly a problem, given that humans have engineered a radically different environment for ourselves over the last 200 years. At the pace of evolution, we were unlikely to evolve in time to adapt.

As epigenetics modifies that understanding, we are now realising that adaptation is occurring and can occur on an infinitely faster timescale: practically in real time, or at least between generations. In the context of equipping our groups to be fit and capable for the future, this couldn't be better news. Looking at the transmissibility of recovery and resilience from trauma gives me hope that we can learn the lessons we need and pass them on to our descendants.

Healing as Collective Intelligence

I have no doubt that groundbreaking research will continue to offer ways to maximise our collective intelligence by strengthening our resilience and honing our pro-social skills. In the meantime though, we already have so many resources that we could be making more of. Old-fashioned skills like the art of conversation, attentive listening and a capacity to be at ease with our emotions and the emotions of other people; these are the building blocks of all collective intelligence. They are the mechanism for establishing collective wisdom, rather than collective trauma, as the legacy we pass on to our descendants. They are dependent not on technology but on time and attention.

We still reach for them instinctively at times of crisis, such as bereavement. The Jewish practice of sitting *shiva* is an example of positive social behaviour and ancestral wisdom that has endured for thousands of years. In the Jewish tradition, when somebody dies, the bereaved family will sit with each other for a week. They open their house to a constant stream of visitors who bring comfort and consolation, jokes and food. They are encouraged to feel their emotions, talk about the person who has died and experience their grief within a community. They're expected to do the same for their friend or neighbour, when that person is in need. The former chief rabbi Jonathan Sacks wrote in his book *Morality*, 'It is a period when you are hardly alone. It is exhausting, but it achieves many things. Above all, it prevents you from retreating into yourself. It softens the jagged edges of grief.'

It's not just *shiva* where the collective is called upon to support individuals through times of distress, of course. Healing circles, group therapy, 12-step recovery meetings: there are many examples of how we seek out community and opportunities to talk and listen to one another. We know, instinctively, that the support of a group can be vital in recovery. This is another example of inherited collective intelligence that neuroscience can shine a light on, confirming what we feel, deep in our bones (or in more neuroscientific terms, as a manifestation of embodied cognition!).

Some of the most robust and significant findings on how to instil greater resilience against the effects of early-years trauma, for example, point to the power of the collective. Anne-Laura's research into resilience in children identifies the child–caregiver relationship as crucial in the early years, but by the teenage years it's friends and peers that can make the biggest difference. Having a supportive group, or even one or two good friends, protects teenagers against mental health problems and keeps them integrated into a wider community. This positive effect lasts well into early adulthood. 'Friendships at age fourteen predict how the brain responds to social rejection at the age of twenty-four,' Anne-Laura told me. 'Your adolescent friendships shape the way your brain responds to social situations even ten years later.' They say it takes a village to raise a child and, neurologically, this seems to be the case. We can all help to provide a buffer against the negative impacts of stress, and in this way maintain the collective intelligence of our groups.

Language is key to many of the ways in which we support one another to heal and grow, using our experiences as the raw material for knowledge that can be of benefit to the group. Sophie Scott is professor of cognitive neuroscience at University College London and studies speech and communication, in particular the use of humour and the positive power of laughter. She and her team have found that people are thirty times more likely to laugh with others than when they are by themselves. Laughter releases endorphins, makes you less susceptible to pain, exercises your body and engages your social brain. It is a tool for social bonding but also a profoundly healing experience for an individual. This perhaps helps to explain why comedy has evolved across cultures. Even in times of great hardship and suffering, dark humour can help people to bond, heal and try to make sense of their experiences. Artistic expression can fulfil a similar function.

But even without laughter or art, it's extraordinary how people can connect if they simply take the time. Twenty years ago, American psychologist Arthur Aron showed that if you put two strangers in a room and ask them to look each other in the eye for four minutes, while asking each other a series of personal questions, they develop a sense of profound understanding of each other. Four minutes of focused attention and honest communication are all it takes to feel bonded to somebody else.

What happens when people struggle to get others to take the time, pay attention and listen to them? Unfortunately, as we discovered in the previous chapter, it can lead to conspiracy theories and extreme tribal beliefs. If individuals feel unheard by friends and family, there is a tendency to put their own narratives on loop inside their own heads, resulting in negative rumination that prevents them from healing, learning or changing their minds. Compassion and pragmatism are the tools for reintegrating a person into a group and working together on a meaningful consensus. In their absence – or worse, if the person is ignored, ridiculed or scapegoated – the result can be alienation, victimhood or even violence.

Feelings of being unlistened to and overlooked have been linked with a low willingness to forgive. This can have devastating consequences. Edith Eger, Holocaust survivor and psychologist, who specialised in the treatment of post-traumatic stress disorder, wrote in her autobiography *The Choice*, 'Suffering is universal but victimhood is optional. There is a difference between victimisation and victimhood. All of us are likely to be victimised at some stage. We will suffer abuse, injury, misfortune or failure. Victimisation comes from outside. Victimhood comes from the inside . . . No one makes you a victim but you.'

She identified victimhood as a way of thinking and being that is rigid, blaming, pessimistic, stuck in the past, unforgiving, punitive and without healthy limits or boundaries. In some tragic scenarios this can lead people to commit acts of revenge against those they perceive as guilty. Even where the results are not violent, Eger points out that this mode of thinking leads to our becoming 'our own jailors'.

Nelson Mandela famously called for forgiveness and reconciliation after the end of apartheid, reminding his fellow countrymen and the world of the concept of *ubuntu*, a Nguni-Bantu word for the common human consciousness that binds all people together. It is sometimes expressed as 'I am because we are'. Or as the Reverend Desmond Tutu put it, '*Ubuntu* is not, "I think therefore I am". It says rather, "I am human because I belong. I participate. I share."'

But Mandela's call for forgiveness, like the Truth and Reconciliation Commission over which Tutu presided, came at the end of a long struggle for justice that eventually succeeded. There could be no reconciliation until the exploitation, at least in its most brutally institutionalised form, had stopped.

Ubuntu has inspired Jean Bosco Niyonzima, the executive director of the Ubuntu Centre for Peace in Kigali, Rwanda, to investigate collective healing from collective suffering. He and his colleagues tackle the trauma arising from the Rwandan genocide of 1994, during which approximately 800,000 people were murdered and an estimated 26 per cent of those who survived developed severe PTSD. In June 2017, the team conducted a first-of-its-kind pilot study into the power of community-based collective healing following genocide. They trained forty community healing assistants from the Kamonyi District, southern Rwanda. Each pair of assistants was assigned to work with fifteen to twenty trauma-affected community members, for fifteen weeks. More than 600 people participated in group-based healing exercises, including meditation, to boost their interoceptive abilities to help them pick up on signals held in their bodies. They were taught how to use this information as a tool to rethink their sense of identity, moving beyond the personal to encompass wider aspects of humanity.

These identity-expansion exercises were combined with communal recovery rituals similar to shamanist practices. People used drumming, chanting, dancing, trancing and metaphors to express their feelings. Storytelling became a collective tool for organising memories into a coherent narrative that allowed each person to express their perspective.

After the fifteen weeks of this support, people's wellness was measured again. The results demonstrated clear reductions in PTSD symptoms including thoughts of suicide, less fear over conflicts to do with ethnic relations and substantially increased feelings of safety and social cohesion. As Niyonzima puts it, 'When society is traumatized, healing must take place collectively.' This collective rehabilitation 'consists of an integrated process of naming collective wounds together, processing them together, and interpreting their meanings together to envision a future free from revenge and violence'.

Forgiveness and reconciliation are among the most cognitively costly activities the human brain can carry out. To recover from the past requires the ability to demolish memory circuits, to make strands of DNA dance contortions to express different genes, to break habits in thinking and replace them with the protective healing process. All this is hard brain work but, as Edith Eger says and Mandela knew, it is necessary work – both for the individual who can release themself from the prison of hatred, and for the group that will otherwise remain trapped in a conflict that makes success impossible.

My hope is that cognitive science can support us to recognise the task of moving on from the effects of collective trauma, in all its complexity and importance. Dr Stefanie Gillson is public psychiatry fellow at Yale University. In 2019, she wrote, 'As clinicians we must acknowledge and appreciate the biological implications of historical trauma. Furthermore, we need to increase our understanding of those who suffer from their ancestors' traumatic experiences. This insight is critical for developing treatment strategies and methods to better target the current impact of past generational harm.'

There are groups of people alive today whose life chances – for good and bad – have been shaped by the behaviours of their ancestors. Money and privilege can be inherited. So can a tendency to psychiatric disorders, incarceration and being marginalised in society. There's a key moment in the podcast *Renegades: Born in the USA* where Bruce Springsteen is talking to Barack Obama about their friendship. When Springsteen reaches the topic of systemic racism and the legacy of trauma it has inflicted on countless people in the USA and around the world, he stumbles and flails. 'Why is it so hard to talk about race? . . .You have to deconstruct the myth of the melting pot – which has never fundamentally been true. Admit that a big part of our history has been plunderous and violent and rigged against people of color. We are ashamed of our collective guilt. We would have to admit, and to grieve for what has been done. We would have to acknowledge our own daily complicity. And to acknowledge our group membership and that we are tied to the history of racism . . . Those are all hard things for people to do.'

They are hard in every sense: morally, philosophically and legally. They are also hard in a very literal sense: at the level of neuronal activity. The cognitive work of tackling abstract concepts that require us to mentally time-travel back into the past and consider events in their own moral context is hard. The cognitive work of thinking ourselves into the future and asking ourselves what we want for our descendants is hard. The cognitive work of thinking beyond our own interest groups and our tribes is hard. It is easier for our brains to try to ignore the evidence from the world around us, but that is not an intelligent strategy. Addressing these issues will take some serious mental gymnastics from us as individuals. There may well be legal challenges on behalf of groups that have been affected, using findings from cognitive scientists as evidence. There is likely to be legislative change on a wider scale.

This might be the ultimate test of collective intelligence, but the potential rewards are huge. If we can learn how to be Good Ancestors by expanding our thinking across generations, we will be doing the cognitive work that ensures we're passing on more collective intelligence and less collective trauma. By adopting cathedral thinking, we can redress the damage of the past and consciously build our futures.

Ultimately, the biology underpinning memory and imagination creates the story we tell about ourselves. It also strongly impacts the stories our descendants will be able to tell. Let's get going on creating the new Notre Dame.

EXERCISE: Be a Time Lord

If your group has a complex problem that you need to approach with longer-term thinking, it can help to get into role before you tackle it. Choose a costume that will denote some members of your group as Time Lords. You don't all need to be wearing the same thing. It could be a ceremonial hat, a gown, butterfly wings, whatever you have to hand so long as it's noticeable enough to remind you of your collective role as guardians and spokespeople for future generations.

Once you're in costume, sit down, close your eyes, and imagine yourself into the role of a Time Lord of the Future.

The first part of the exercise is to dedicate some solo thinking time to the problem. What will it look like five years from now, ten years, thirty years and beyond? Try to imagine your own needs in the future, and those of the people around you. Make notes on concerns and ideas.

Then come together as a group to share your thoughts. Deploy some of the techniques we've discussed, such as active listening and brainwriting, to explore problems and solutions with the long term in mind.

CHAPTER TEN

Melding Minds: Us and AI

How about we all become cyborgs? With the best will in the world, cathedrals to collective intelligence are going to take some building. We need all minds on deck, now, to tackle global problems and build ambitious solutions. So let's enlist *artificial* intelligence as well as humanity's cognitive capacity. Maybe we could implant prosthetic devices in our brains to boost our memory storage or our emotional intelligence. Even better, let's create a super-brain by wiring up a network of human brains and combining them with artificial computational power.

Neuro-engineering is making exciting advances, but for many of us this science can feel more like science fiction. It's hard to know what's useful and helpful and what's not. In this chapter I will be demystifying the work that's under way to meld our minds with others'. I'll also look at the need to design resilience and flexibility into these long-term projects so that they become the embodiment of cathedral thinking. There are ethical as well as practical implications that we can't afford to ignore, so I'll be addressing those, all from a starting point of excitement. We are living in a golden age of scientific exploration. The landscape we're exploring is that of consciousness itself, and what we find will shape the next stage in the evolution of collective intelligence.

We already live in an exploding age of artificial intelligence, of course. It's all around us: driverless cars are on the roads, Siri's on our phones and care-bots are monitoring patients' health. Computational systems can beat chess champions, win poetry competitions, obliterate ping-pong professionals with high-precision swipes and even wage war on our sense of reality by creating fake news in such abundance that it swamps the authentic posts that human beings offer.

In the hospital setting, computer programs can now detect some forms of cancerous tissue with higher sensitivity, greater accuracy and more speed than highly trained radiographers. Many aspects of mental health can be managed online, by systems that monitor how we're feeling and can provide personalised support and guidance to boost our resilience when we need it. In schools and colleges, online courses are offered with tailored homework that's set automatically for each student, and educational apps are becoming embedded within the curriculum. Drones are starting to replace couriers, and flight paths are controlled by robot traffic controllers.

The benefits to society that AI brings are obvious. Computers are supremely skilled at the many tasks we set them. We have, after all, designed them to be that way. Machine learning, the process that underpins AI, was inspired by and modelled on the neural networks found in the human brain. Systems take in information from the environment and use it to first build a framework of reality, then learn from past experiences, gain wisdom and direct their response. The latest AI systems can even process, integrate and build on information far more rapidly and accurately than humans.

But as we've also seen throughout the book, social and emotional intelligence are the skillsets that underpin successful collaboration, and they are beyond the current capacities of artificial intelligence. Remember those warring Wikipedia bots? Without a theory of mind to support their empathy and a sense of agency to support their personal responsibility, they were incapable of organising themselves to work together effectively.

There are many ways in which the human brain remains – for now – vastly more flexible, effective and creative than AI. The question that none of us can answer is how long it will take AI to evolve these skills for itself. All we can say is that it will be a great deal quicker than the hundreds of thousands of years it took human beings.

Can we, and should we, treat AI as simply another form of diverse intelligence to bring into the mix? Perhaps. One example of how this is already playing out is in the task of caring for our ageing and increasingly fragile population. Dementia is a scourge of our times and demands huge reserves of patience and empathetic interaction from carers if sufferers are to be well looked after. In 2014, I interviewed Goldie Nejat for a BBC radio broadcast. She's director of the Institute for Robotics and Mechatronics at Toronto University, which looks at novel ways we can support the world's elderly population as their cognitive abilities decline. She's interested in how robots can help jog memories and support people in their daily activities while also keeping them socially active. She introduced me to three robots she's created. There's Brian: he encourages those with dementia to eat; Casper, who prepares food; and Tangi, who helps people get together for bingo. 'We're trying to design socially assistive robots that can engage people in recreational activities, memory games, as well as help them with activities of daily living that they might find difficult on their own,' Nejat told me.

Nobody wants to see a situation in which bots replace human carers completely. The embodied cognition, the theory of mind and the ability to realise that what somebody needs most is a hug are all attributes of humanity's sophisticated social and emotional intelligence that are literally irreplaceable. But care-bots *can* take on many interactions, and so free up the human carer for others. And with each robot costing £4,000 to £8,000, it's significantly cheaper than a human healthcare assistant.

This form of collaboration between human beings and AI is built on sharing workload according to skillset; it doesn't involve fitting the brain with a computer chip or wiring brains together, but neuro-engineering of that kind is already an advanced science. There are hundreds of thousands of individuals across the world benefiting from neural implants: prosthetic devices embedded in their brains that can, with the flick of a switch, turn off the symptoms of Parkinson's disease or addiction. My neighbour, Kate, for example, has such an implant to help her manage her Parkinson's. She's still out cheerfully jogging each morning thanks to a device that stimulates specific brain circuits with electrical currents, switching off her symptoms of tremors and depression. Transhumanism – the belief that humans can evolve beyond our physical or mental limitations with the help of technology – is already a lived reality for people like Kate.

There are clear life-changing benefits to this technology for people with medical conditions, from paralysis to OCD. But as our technological capacity increases, it seems inevitable that we will start to wonder whether people without life-limiting conditions might also benefit from a little neuro-engineering to boost their cognitive capacities. We know that the lure of greater intelligence is powerful, as the use of smart drugs, microdoses of psychedelics and even illegal gene-editing of embryos has shown. If our society is going to embrace transhumanism further in the future, surely it makes sense to aspire to do so in a collaborative, collective way rather than simply adding to an individual's skillset. I'm thinking about ramping up our collective intelligence by literally bringing mankind's minds together, fusing our cognitive capacity in order to create a super-brain cloud aided by artificial intelligence. This would involve wiring up human brains to create a living hive mind. It might sound like the stuff of science-fiction but it is within plausible reach.

There are so many questions to answer about our path towards man–machine interfaces. Would the merger of human minds with computers *actually* help boost our collective intelligence, or would we be compromised by the undermining of our pro-social behaviours? Would AI remain our partner or would it turn itself into our overlord, harvesting our mental energies or controlling our thinking? Would we, or AI, start to have difficulties differentiating between them and us, causing dilemmas around identity? What about privacy, hacking and moral responsibility if we're all wired up to one another via computers? The practical, philosophical and ethical considerations are vast; the issues we as a society have to grapple with in order to keep up with the pace of change are deep.

It's important to embrace uncertainty and engage our slower, less emotional thinking when we consider these questions. They are cognitively challenging and there are no easy answers, as Stephen Hawking acknowledged when he talked about his ambivalence on the subject. Hawking used machine-learning technology in order to circumnavigate his severe motor neurone disease, which denied him independent speech and movement. The last iteration of the AI that he used was similar to an exceedingly advanced form of autocorrect predictive text. It memorised Hawking's previous communications in order to suggest words he might want to use next. Despite benefiting hugely from AI and being largely an advocate for it, Hawking never hid his misgivings, saying to the BBC in 2014 that its 'full development could spell the end of the human race'.

Similarly, Elon Musk has been investing in neuro-engineering for years but warns that we may fall so far behind AI that we end up as little more than diverting pets for our robot owners. 'I don't love the idea of being a house cat,' he said in 2016, 'but what's the solution? . . . I think maybe the best [thing for humans to do] is to add an AI layer.'

If even the very well informed and personally implicated are ambivalent, it's surely understandable that many of us are nervous. Let's try to resist fear, though. (As we know, it makes us stupid!)

After all, every technological revolution has made people jittery at the time; and there are always losers as well as winners when things change, but that doesn't mean that change is inherently bad. Do we look back at history and wish the Industrial Revolution hadn't happened? Do we envy other European nations who were slower to industrialise, and wish we'd let them take the lead? Do we look at the Luddite uprising, and wish it had been a success? Do we remember the 1970s, and wish those pesky word processors and spreadsheets had been outlawed in order to save clerical jobs?

I believe in the possibilities for progress offered by science but I also believe that we can't afford to trust that such progress is assured, or that its benefits will be shared out fairly. That only happens when many people from different walks of life push for it. And for that collective push to occur, we all need to understand what's possible and what's at stake. The next tech revolution is on its way and has the potential to be wider and yet more intimate in its reach than any before. We cannot afford not to think about it. We need to remember the lessons of cathedral thinking and draw on those uniquely human skills of faith, trust and imagination. We need to be optimistic.

Maybe the example of Garry Kasparov can inspire us. He is quite possibly the greatest chess player who has ever lived. He spent his youth polishing the fine art of creative strategy until, in 1985, at the age of twenty-two, he became the youngest ever undisputed world chess champion. He held this title pretty much consecutively until his retirement, decades later. The exception was in 1997, when some whippersnapper entered the arena and, in a highly publicised match, Kasparov became the first world champion to lose to them. Who was this awesome competitor? The IBM supercomputer, Deep Blue.

Kasparov didn't give up the game after being beaten by a computer. Rather than being dejected he felt inspired, and the next year he returned to competitive play having devised a new format: advanced or cyborg chess. This novel method brings together man and machine, combining their forces rather than pitting them against each other. It merges human creativity, tactical play and strategic planning with high-acuity error-spotting and brute computing force to systematically generate and test all possible solutions. Cyborg chess can evaluate 200 million positions per second yet produce sophisticated and creative games. Today, any old chess computer programme can clean up, even against a grandmaster, but pairing an average player with an average computer creates a winning combination that can wipe the floor with even the best supercomputer. In 2017, Kasparov wrote a book called *Deep Thinking: Where Machine Intelligence Ends and Human Creativity Begins*. 'We must press forward ambitiously in the one area robots cannot compete with humans: in dreaming big dreams. Our machines will help us achieve them.'

What Can We Do Already?

As we've seen, neuro-engineering is now being used to treat a wide variety of health problems by delivering highly targeted electrical currents to particular brain circuits. Deep-brain stimulation, via a neural implant, was first deployed for the treatment of Parkinson's disease in 1997. Similar devices now help with addiction, OCD, food disorders and anxiety.

Recently, scientists have developed even more sophisticated implants called 'neural dust', which are tiny wireless sensors. They use ultrasound to both monitor and stimulate particular brain cells, and communicate with an interrogator unit that's fitted between the layers of membrane that wrap around the brain. The scale of neural dust – each particle is measured not in millimetres but in nanometres – and the fact that they don't give off heat means that they don't cause scarring or inflammation. Thousands of them are used in each intervention, acting as an 'electroceutical' to treat conditions from sleep apnoea to epilepsy. Other brain–computer interfaces (BCIs) help amputees, paraplegics and people who have reduced movement after strokes, while retinal implant systems can help people who were previously blind to see again.

The ambition and increasing effectiveness of neuro-engineering takes my breath away. Artificial synapses and external memory hard-drives are already in development and these projects might eventually deliver an effective treatment for dementia that can reverse the damage to people's memories.

Elon Musk's high-profile company Neuralink is working on neural lace: an injectable network of silicon particles that would contribute a digital layer to the brain. Musk has been flamboyant about its purpose, claiming in April 2022 that it will produce a medical wonder-tool that can cure paralysis and also 'solve a very wide range of brain injuries [and health problems] including severe depression, morbid obesity, potentially schizophrenia – a lot of things that cause great stress to people'.

Whether the breakthroughs are two, ten or twenty years away, there is every reason to believe that neuro-engineering will continue to deliver life-changing solutions for people whose cognitive abilities have been impaired. But will it be limited to those with a medical need? As reported by the BBC, Musk's long-term ambition seems to be to usher in an age of 'superhuman cognition', in part to combat the threat of an artificial intelligence so powerful that he believes it could destroy, or become the overlord to, the human race. But will AI really be able to supersede those traits that we think of as typically human, those abilities that underlie our capability for collective intelligence?

Well, it can already emulate them. Scientists have started to engineer emotional intelligence, once again for use by people with a medical diagnosis. Peter Robinson, professor of computer technology at Cambridge University, is busy creating 'socially and emotionally adept technologies' that attempt to infer people's mental states from facial expressions, vocal nuances, body posture and gesture. He claims his computer can be as accurate as the top 6 per cent of people at this. Peter is developing wearable systems to help people with autism spectrum conditions and Asperger's Syndrome boost their emotional-social understanding and ability to pick up on what others are thinking.

Few would argue that helping somebody with a diagnosis of severe autism is inherently worthwhile but the same tech is already being used beyond the clinical field, which makes it an interesting test case for the kinds of questions that we will increasingly have to grapple with. It has made its way into key aspects of our social fabric, including the judicial and legal system and the healthcare, education and employment sectors. And here the balance between benefits and risks to individuals and the group is less clear-cut. A company called HireVue, for example, sells AI-driven video-based tools that can recommend which candidates a company should interview, based on the data they harvest from facial expressions during preliminary conversations. Oxygen Forensics allegedly offers emotion-detecting software to police. A company called Cogito supplies voice analysis carried out by algorithms for use by staff in call centres, to pick up on when customers are becoming distressed.

At the end of 2019 the research centre AI Now reported that the sector, which was estimated then to be worth $20bn (£15.3bn), was growing fast. Professor Kate Crawford, AI Now co-founder, wrote that the technology: 'Claims to read, if you will, our inner emotional states by interpreting the micro-expressions on our face, the tone of our voice or even the way that we walk . . . It's being used everywhere, from hiring the perfect employee to assessing patient pain, to tracking which students seem to be paying attention in class.' She voiced concerns over its reliability, calling for new laws to restrict its use. 'At the same time as these technologies are being rolled out, studies are showing that there is . . . no substantial evidence for a consistent relationship between the emotion that you are feeling and the way that your face looks.'

As we saw in Chapter Seven, our facial expressions do not necessarily reflect our emotions and they are certainly not a global language, more like local dialects varying from region to region. Current AI technologies may well not pick up on these variations, which means that the use of this technology, as it stands, is more likely to amplify our bias and narrow our collective intelligence than expand it.

So far there has been barely any more discussion of the issue, let alone legislation. I'm no fan of rushing to legislate around technology but I do believe that we the public, policymakers and experts need to have a conversation about this and the other applications of artificial intelligence that we're examining in this chapter.

Creating the Brain Cloud: Our Next Frontier

What we can already do is extraordinary, and so much of it is astoundingly positive. Cyborgs are a reality. People with degenerative conditions like motor neurone disease can be helped to move and communicate. A paraplegic volunteer named Juliano Pinto kicked the opening ball at the football World Cup in Rio thanks to a brain–computer interface and a robotic exoskeleton. These advances are all impressive and moving examples of collective intelligence at work. Neuroscientist Dr Miguel Nicolelis, the lead scientist on the Walk Again Project that delivered Pinto's kick, congratulated its 150 researchers afterwards, saying it was 'a great team effort. It was up to Juliano to wear the exoskeleton but all of them made that shot. It was a big score by these people and by our science.'

What comes next promises us even more. Although your average neural engineer may not have the vast platform and profile of Musk, they have been beavering away on connecting one brain directly to another for the last decade or so, with quiet success.

Dr Nicolelis has carried out some of the pioneering work in this field. In 2015, he and his colleagues at Duke University published two papers on their research into direct brain-to-brain interfaces, which they call 'Brainets'. In one study they fitted neural implants into the brains of three macaque monkeys and then taught them all, independently, how to move a robotic arm displayed on a screen simply by imagining doing so. They then taught the monkeys to work as a team to move the arm in certain ways, such as reaching for a ball. Each monkey had the ability to control a particular range of movement and no one monkey could pick up the ball alone. At least two had to be involved in order to complete the task. Even though the monkeys' brains weren't directly connected one to another, they collaborated and figured out how to complete the task together.

The same team also created a Brainet of four adult rats that did wire the animals' brains up directly to one another. The network successfully exchanged, processed and stored information, essentially creating an organic computing device as the rats interacted with each other. They were able to solve various problems together, including image processing and storage and retrieval of information, and all the participants became more accurate and faster in their responses. Nicolelis believes this principle could eventually be applied to humans and 'extended indefinitely, to enlist millions of brains to work together in a "biological computer" that tackle[s] questions that could not be posed, or answered, in binary form'.

In 2019, Chantel Prat and other scientists at the University of Washington, Seattle took a significant step towards this. They created a Brainet that they described as 'the first multi-person non-invasive direct brain-to-brain interface for collaborative problem-solving'. The group of three people was asked to solve a simple task (figuring out how to rotate a block on a Tetris-style game screen). Two of the three participants were designated as senders of information. They could see the screen and had to decide how to manipulate the block in order to move it into the right space. Their brain signals were recorded via EEG caps and sent to the third person via transcranial stimulation. The receiver could not see the screen but had to interpret the information and implement the decisions. The group achieved on average more than 80 per cent accuracy. Like the monkeys before them, they were collaborating as a group to move an object on a screen, but this time their brains were communicating directly via electrical signals rather than synchronising through learned behaviour.

Prat and her team wanted to see whether this brain-to-brain network could be made to function like other networks, in which people learn to trust and rely on some people's inputs more than others. So they designed a way to make one sender's signals less reliable, by making them noisier and harder to pick up. The sender with 'better' information was preferred by the receiver, who learned how to tune out the noisy signal in order to arrive at a correct decision about how and when to rotate the block. This finding might eventually be developed to function like a dimmer switch, enabling someone to turn down the volume on people not contributing effectively to a group.

The researchers also used the Brainet to play some quite successful games of Twenty Questions, with ten different people hooked up to one another. This all points the way, as they put it, to 'future brain-to-brain interfaces that enable cooperative problem-solving by humans using a "social network" of connected brains.' We're seeing the foundations of work that could ultimately lead to a human computer that could be harnessed to solve problems way beyond the limitations of our individual intelligence or current computational powers.

In the meantime, we're making strides towards being able to transfer specific information from one brain to another via electrical stimulation rather than language. Previously this has always felt out of reach. Neuroscientists have been able to determine that activity is occurring in a particular region and to associate it with different functions such as motor control, pleasure, memory, even highly abstract functions such as empathy and guilt. But they have not been able to say anything about the content of that activity. We could say that learning was taking place when we saw the hippocampus lighting up, but could not say anything about the content of that learning. Until, in 2013, a precise mechanism for how the hippocampus stores a specific piece of information was observed in action, taking place in real time as a rat learned how to complete a task.

This experiment, quite simply, blows my mind. Sam Deadwyler of Wake Forest School of Medicine, Winston-Salem, and his team, were studying rats carrying out a short-term memory task when they observed very specific patterns of activity within ensembles of nerve cells in the rats' hippocampi. They recorded the electrical dances of synapses firing in the precise configuration that represented the specific event being encoded. For me, watching the film of the demo is a little bit like watching the Enigma Code being cracked.

Even more mind-blowing, in a follow-up study the electrical 'representation' of the task was extracted from a well-trained 'donor' animal that had successfully learned how to perform the memory task, and was then delivered via brain stimulation to a naive 'recipient' animal who had never been exposed to the task. By transferring the donor animal's hippocampal firing patterns via stimulation, the recipient rat brain acquired the memory of how to carry out the task. The ability had been induced in the brain through electrical stimulation and passed from donor to recipient without language, imitation or ancestral genetic code. It begs the question – might it be possible to take the neural information from one brain and use it to induce, recover, or enhance memory-related processing in the brain of another subject? The ramifications of this research are profound. It's one firm, weighty step towards uploading the human brain and merging minds in a cloud.

Along similar lines, Theodore Berger at the University of Southern California is creating a microchip-based prosthesis for the hippocampus that supports its work to turn impressions into long-term memories. If this development could eventually be paired with neural lace to provide neuronal scaffolding, we would be much closer to being able to connect human brains in real time.

Entering an Ethical Quagmire

These advances and the many others that are taking place all over the world are impressive. Even if the scale of tasks that Brainets can currently tackle is very small, the technology will develop to allow those tasks to get bigger. And as the invention of neural dust shows, challenges such as how to place foreign objects in the brain without causing scarring can be overcome. So these already impressive advances are set to get ever more so, which is hugely exciting. And terrifying.

For now, the act of linking brains together only occurs in highly controlled laboratory environments among volunteers, but once it becomes easy to do – whenever that is – how might that change? Who would be able to join the Brainet? How would we ensure that participation was neither coerced nor denied? If you had to have a certain IQ result to be eligible to join the brain cloud, for example, and life outside the cloud made you a second-class citizen, we would truly be in dark territory.

Recent history is littered with horrendous examples of what can happen when pseudoscientific thinking about intelligence is used to justify the unjustifiable. The apex of this horror is the Nazis' attempt to wipe out Jews, homosexuals, Roma, Black and Asian people, disabled people and various other groups who were proclaimed to be undermining the racial purity and intellectual vigour of the so-called 'Aryan race'. There have been other less violent but still highly coercive and immoral instances of eugenics based on notions of intelligence. In Sweden, people with learning difficulties were subjected to forced sterilisation until well into the mid-1970s. Some 60,000 people were treated in this way. That such acts grew out of science provides a stark warning and should push us to engage with the ethical implications of these technologies.

Beyond these questions of the right to access and the right to refuse, what happens to the privacy of our own minds if there's potential for our brains to be hacked? What about mind control? Individuals could intentionally manipulate the neural signals being sent to recipients, changing their thoughts not through persuasion or even propaganda but via a direct, uncontrollable route. And if we can merge minds, would that result in irreversible cognitive changes? How might this alter human cognitive – and possibly moral – capacity?

One troublesome aspect of all this is that much of the work is being undertaken by the technology giants of the digital communications sector, who are developing their own neurotechnology. These commercial companies exist to make profit for shareholders and have already shown themselves to be indifferent to ethical questions and the damage done to users' mental health.

As well as Elon Musk and Neuralink, Mark Zuckerberg has been investing in technological telepathy. In the summer of

2019, scientists funded by Facebook and based in San Francisco published a paper on their development of a headset that could transfer a person's thoughts directly onto a computer screen. For now it can only decode a small selection of words but the firm hopes it will one day work seamlessly to transcribe thoughts and transmit them into the brains of other users. Group work would take place not so much online as in-brain. This raises profound questions about privacy, as well as who owns the intellectual property in thoughts generated in brain-to-brain interfacing. And who is responsible if the group does something criminal?

Frank Vetere, director of the Microsoft Research Centre for Social Natural User Interfaces at the University of Melbourne, researches human–computer interactions and artificial intelligence technologies that improve social wellbeing. Although he's excited about the possibilities, he's also concerned. Industry and government authorities already collect vast interrelated data sets crammed with personal information, which gives them a huge amount of power over us. 'Insurance companies now collate health data and track driving behaviours to personalise insurance fees. Law enforcement uses driver licence photos to identify potential criminals, and shopping centres analyse people's facial features to better target advertising.'

But the data is only the start of the problem. Vetere is sceptical about the likelihood of all such data being accurate; but even if it is, the way that it's interpreted and applied is subject to flaws and errors, assumptions and blind spots. As we've discovered, humans can make serious errors, even when in a group. AI might not always pick up on these errors; it might in fact amplify them. An algorithm can be as biased as the person who wrote it. And as Vetere puts it, 'the opacity of AI processes make it difficult to redress algorithmic bias'. His warning is stark: 'Some AI systems are sexist, racist, or discriminate against the poor.'

Imagine that the presumptions of a biased AI are being shared with someone like your insurer or future employer. Now ask yourself where this might go if your brain waves, memories and cognitive capabilities could also be uploaded. What conclusions might be drawn and predictions be made as the technology becomes ever more intrusive? And how might they shape your life or restrict the choices available to you? Would your individual autonomy be destroyed by sharing access to your brain – the intimate seat of your selfhood?

It was in response to these kinds of questions that in September 2019, Britain's Royal Society called for a government inquiry into the use of brain–machine interfaces. Their report *iHuman: Blurring Lines Between Mind and Machine* highlighted the ethical concerns surrounding 'enabling brain signals, and ultimately thoughts, to be detected or stimulated by external devices' in the future.

Co-chair Professor Christofer Toumazou from Imperial College London underlined the fact that while the technology is still mostly experimental, its effects are destined to be huge. 'By 2040 neural interfaces are likely to be an established option to enable people to walk after paralysis and tackle treatment-resistant depression; they may even have made treating Alzheimer's disease a reality. While advances like seamless brain-to-computer communication seem a much more distant possibility . . . The applications for neural interfaces are as unimaginable today as the smartphone was a few decades ago. They could bring huge economic benefits to the UK and transform sectors like the NHS, public health and social care, but if developments are dictated by a handful of companies, then less commercial applications could be sidelined.'

To me, the report reads like an argument in favour of treating brain–machine interfaces as one of humanity's most pressing and exciting cathedral-thinking projects. The job of figuring out how to develop and deploy this technology is vast, and requires a wide conversation among a really diverse and representative selection of people. We need to hear from everyone on this issue, from experts to end users, and to listen to each other's hopes and fears.

As Dr Tim Constandinou, another of the co-chairs, put it, 'We should act now to ensure our ethical and regulatory safeguards are flexible enough for any future development. In this way we can guarantee these emerging technologies are implemented safely and for the benefit of humanity . . . [We need] a national investigation to identify the UK's priorities and let the public help shape how the technology develops and where we want it to take us.'

Melding Minds to Dream Much Bigger

Perhaps the merging of our minds with AI is the moment we find ourselves reaching beyond human parameters, overthrowing natural evolution and veering dangerously towards hubris. It has that potential. But I see no reason to succumb to panic about AI plunging us into a dystopian nightmare. There are global regulatory checks already in place on what we can do with this technology and as the conversation around the issues takes off, I believe we will be able to successfully navigate the ethical quandaries. There's no room for complacency, clearly, but I suspect our collective experience with social media has made us more aware of the need to regulate *alongside* such world-changing digital technology rather than in its wake. Hopefully we'll be wiser this time. And ultimately, nobody wants a device implanted into their brain that can read their thoughts and control their behaviour. We have a collective aversion to that scenario that will protect us, so long as we remain vigilant.

Meanwhile, the possibilities opening up could spur our imaginations to ever greater heights. Our tech-enabled capacity to experience how the world looks and feels from inside the mind of another person is one of many tools that could expand our 'we' thinking. Pawel Tacikowski, a postdoctoral researcher at Karolinska Institutet in Sweden, has been investigating the neuroscience of how we maintain a consistent and unified selfhood, as well as how we relate to other people. He told an interviewer from *Science Daily* that as a child he had always wondered whether he was the same person when he woke up as he was when he went to sleep, and had simply never grown out of asking these sorts of questions. He wanted to investigate how much influence the physical body has on our brains' concept of self so he developed a perceptual illusion experiment, where pairs of friends 'swapped bodies' using virtual-reality tech. Tacikowski found that people's beliefs and personalities very quickly altered and began to mirror what they knew of their friends'. The team from the Brain, Body and Self Laboratory, led by Henrik Ehrsson, outfitted pairs of friends with goggles showing live feeds of the other person's body from a first-person perspective. To further the illusion, they applied simultaneous touches to both participants on corresponding body parts so they could also feel what they saw in the goggles. After just a few moments, the illusion generally worked so effectively that when the researchers threatened the other's body with a prop knife, participants broke out in a sweat as if they were the one being threatened. The perceptual body-swap only lasted for a brief period of time but it was long enough to significantly alter participants' self-perception. Before the experiment, people rated both themselves and their friend on traits such as talkativeness, cheerfulness, independence and confidence. They were then asked to rate themselves again during the body-swap. Compared to their baseline results, participants tended to rate themselves as more similar to the friend whose body they were 'in'.

I'm fascinated by this study – potentially it provides a new tool with which we can promote feelings of empathy, so that people can share their experiences and belief systems simply by entering a virtual-reality laboratory.

Or what about a Brainet that enables people to share emotions directly, infecting each other with calm and boosting their creativity? A team of researchers at Maynooth University, Ireland, led by Richard Roche, created a network of three people wearing EEG headsets. As the first person relaxed, the slow, rhythmical electrical oscillation of alpha waves, linked to calmness and creativity, increased across their brain. The EEG recording of this brain activity was then converted into a different sensory modality – so either a sound (like raindrops falling), a sight (like a calming image of the ocean) or a tactile stimulation (a relaxing touch) that was presented to the next person in the loop. This, in turn, induced their alpha waves to increase, activating the next relaxing signal to be passed to the next person. Three people become one mass organism responding, relaxing and creating together in real time. The demo (which you can view here: https://vimeo.com/237065016) poses the question: where does the signal, and the individual, begin and where does it end?

Artists Karen Lancel and Herman Maat, who are based at the University of Delft, Netherlands, used the same techniques to deconstruct a kiss and turn it into an immersive installation. Visitors are invited to have their brain waves read or their heartbeat recorded as they embrace. This information is then reconstructed to create a new, digital synaesthetic kiss as the data is translated into a musical score, creating a symphony of the experience. Their most recent project, Empathy Ecologies, extends this exploration of intimate communication even further, across species, using data to make the relationship between humans and plants audible. The kissing pair is surrounded by plants that are 'listening to' the music created by their caress. The plants' fluctuating consumption of CO_2 is in turn monitored and is transformed into a musical score that is added in real time to create an orchestral work of human–plant–technology inter-action. A collective of vastly different intelligences, working together to create something beautiful. The artwork is available to view at https://vimeo.com/422895498.

As we uncover more of the biological underpinnings for abstract thought, creativity, imagination, perception, emotions and reasoning, and develop the technology to explore them, is it too much of a stretch to think we might one day be able to experience the lived intelligence of different species, or recreate great minds long dead by imprinting their thought processes? If we let our imaginations wander, what cathedrals can we envision?

Let's dream what this could look like ... Maybe we could upload brain capacities to access a wealth of past and present thinking. I imagine a scenario in which a vast brain cloud of different people's cognitive powers can be readily accessed, linking up insights and creativity in a majestic way.

If I close my eyes, my imagination turns first of all to how the spaces I live in could be reshaped for maximum beauty, utility and sustainability. In the architectural sphere, the monu-ments and community spaces could be created by combining the engineering ability of Brunel and the artistic inspirations of Gaudí and be overseen by the urban planner and green architect, Cheong Koon Hean. I would add a sprinkle of Frida Kahlo for interior design, snatches of Einstein and Hawking to create spaces that extend into different dimensions, expanding across time and space, and a dash of Marina Abramović to explore how our individual consciousness interacts with others'. I'd add music to enhance the experience – the cognitive resources of David Bowie, Beethoven, Ella Fitzgerald could be pooled to create scores that accompany and heighten the theatrics of the space.

It's just a dream, for now, but imagine if we really could meld minds and position collective intelligence centrally in every aspect of our lives. The world could be altered by the power of our minds to tackle everything, from transport that doesn't cost the Earth to reforming the treatment of witnesses in the courts and designing education that embeds the value of every individual's contribution.

We don't need to wait for technology to do this work for us. If we take a step back and reflect, we can see that simply sitting down and really sharing, really listening, properly communicating with each other, could have the same effect. If we all cultivate the skills that support empathy and compassion, and invest in the value of Us rather than defend the value of Me, we can unleash a torrent of imagination, problem-solving and innovation. And would this not be a simpler, less risky route than relying on technology to solve our problems? I am excited by a future that continues to explore what technology can offer but I do not believe that it could or should offer us an escape route from dealing with our challenges. If we duck our responsibility to step up, we risk sacrificing our humanity.

Neuroscience, like every manifestation of human intelligence, is evolving. In 1952, the Spanish scientist José Delgado, who pioneered the field of brain–machine interfaces, implanted electrode arrays into a bull's brain and entered the bullring to taunt it. When the animal began to charge, he pressed a button to send a radio signal to an implant, which passed an electrical current into its brain. The bull ground to a halt, mid-stride. It was a theatrics of bad taste, designed to massage the ego of the controller and appeal to a crowd that was hungry for the next experience to consume.

Delgado dedicated the remainder of his career to blurring the line between brain and machine. His work focused more and more on therapeutic applications in the treatment of psychiatric patients and his aspiration for technology to support a more compassionate society. In 2005, by which time he was in his late eighties, he spoke with *Scientific American* magazine. He maintained that technology 'has two sides, for good and for bad', and we should do what we can to 'avoid the adverse consequences'.

But human nature, Delgado asserts, is not static but 'dynamic', constantly changing as a result of our compulsive self-exploration. 'Can you avoid knowledge? You cannot! Can you avoid technology? You cannot!'

Delgado died in 2011, shortly before the exponential take-off in his field. I wonder how he would have reflected on the last decade of developments . . . I'd like to think that if he had been able, he would have contributed his wisdom to a conversation with the experimental innovators of today, the Musks and Zuckerbergs. After all, the man who once used brain–machine interface technology to stop a charging bull in its tracks had gone on to experience the reward of putting his intelligence to very different use. He would surely have had insights to offer. Perhaps Delgado would have asked the tech pioneers, and us, to think hard about the direction in which we're taking this work and the impact it might have on humanity's future.

EXERCISE: Create Your Own Brainet

If you have a tricky problem to solve, channel other perspectives by mentally assembling a group of people whom you know well. Imagine having a conversation with them, talking through the issues. What would they advise? Is there anything you could take away from their way of thinking? If they are still alive and available, could you contact them for their thoughts?

Now try to think of someone whom you know and respect but whose perspective is very different from yours. What would be their take on the situation? How would they tackle it? If you would not follow their suggestions, why not? Put yourself in the position of looking for the positive or useful in an approach you don't normally take.

Once you've practised carrying this out as a thought experiment, consider whether you feel able to ask for advice from this person whose thinking is different from yours. Would they be interested in swapping ideas with you? Invite them into a dialogue.

Creating our own Brainets through active listening and respectful conversation fosters collective intelligence and can mean the difference between cracking a problem and giving up.

Epilogue

One afternoon in January 2022, not long after international borders had reopened, travel resumed and Max and I had made it back to the UK from Australia, I perched out on the balcony of my Cambridge flat to soak up the startling contrasts of a wintry sunset. A cascade of pink and orange hues lit up the spires and rooftops as the sun sank lower, growing larger and casting longer and longer shadows across the grey slates, until the final ping as it disappeared below the horizon. I clutched a mug of tea to warm my hands and the twilight settled in as the first stars pinpricked the darkening sky. Over the course of half an hour, one day of my life closed. The night took over. I imagined the coming day and the sun that would reappear on the other side of the building at dawn. My mind painted a picture of its movement across the vast sky, sinking and rising around the pivot point of my gaze.

Except, of course, that wasn't what was happening at all. The sun's trajectory is just a story we tell ourselves. Our logical brains are aware that its setting and rising is a beautiful illusion and that, in reality, as science informs us, it is we, on our planet, that are in motion. The end of our day occurs when the spot we happen to be standing on spins away from the sun's illumination, to return twelve hours or so later, once we have completed a full pirouette. But for each of us dotted across the globe, this illusion of the sun tracking across our stationary sky and setting on our horizon is so powerful that we hardly ever question it. Our brains form an inaccurate perception based on our lived perspective and create a totally convincing illusion centred on our self. This illusion is difficult to shake even once we know how the trick is performed.

It takes mental effort to zoom out from our spot on the Earth's surface and see reality from a wider angle. We cannot hope to keep up that kind of cognitive effort alone. It requires many of us, all sharing the view from our specific vantage points, to create a fuller, brighter, more accurate picture. In order to really see any situation, we have to cast our imaginations beyond our individual self and comprehend that we are simply tiny parts of a much larger interconnected system.

Some societies have never forgotten how to do this. The collective intelligence they have at their disposal is both ancient and fresh as the new day. Aboriginal Australian cultures are vastly different in their thinking from the Western, post-industrial society I've grown up in. The relationships between an individual person and the collective, between humans and other life forms, and between past, present and future are just some of the many points at which these groups of people see the world differently from me. Their thinking about people's collective cognitive abilities can shed light on how humanity might expand its scope of thought.

Indigenous Australian cultures have developed a communal form of tranasctive memory that spans huge numbers of people over many generations. Forget the boost to cognitive power offered by a pair bond: think 50,000 years of wisdom passed from person to person, with nothing written down! Indigenous Australians don't expect any individual to be able to remember everything. They regard everyone as a co-creator of memory and meaning. They share perspectives and create narratives as a group, telling a story that evolves as the collective develops it. Learning is embedded within this process. So if you were learning butterfly names, for example, you might go for a walk round a park or garden with other people and develop a story about the butterflies, attaching them to specific locations. It's not dissimilar to the technique of making a memory palace, in which you remember information by envisaging it as objects in different rooms.

Tyson Yunkaporta is senior lecturer in Indigenous Knowledges at Deakin University in Melbourne, a member of the Apalech Clan in far north Queensland and author of *Sand Talk: How Indigenous Thinking Can Save the World*. He and a team of researchers at Melbourne University set out to investigate the effectiveness of various techniques, ancient and more modern, for boosting an individual's memory. They divided volunteers into three groups and tested them all to get a baseline view of their capacity to recall information. One group was then offered memory training based on the traditional practices, which required them to work together. Another was trained using the memory palace technique, which they used alone, and the third group received no specific training. They were all then tested again. Results improved by, on average, triple in the group that used the indigenous technique, double in the mind palace group and by only 50 per cent in the untrained group. Embedding the facts within a narrative that had been communally created seemed to be key to the success of those trained in the traditional technique.

These results were so significant that the study's authors suggested that medical students, who have to learn a vast amount of information by rote, be taught the Aboriginal memory practice as a way to boost their retention of information.

In the West we regard medicine as a highly competitive field of study that requires an individual to achieve the highest grades across a number of rigorous exams. I find it poignant and hopeful that we're now imagining ways to teach elements of medicine by tapping into the cognitive techniques and embodied wisdom of a more collaborative mindset. Our thinking about thinking evolves through contact with different points of view.

Memory is not merely a tool for passing exams, of course. Our ability to recall our unique experiences and turn those recollections into the story of our lives is the foundation of our individual identity. And yet, Yunkaporta's research suggests that we make memories more efficiently and lastingly as a group than we do on our own. This takes me back to a time when I was carrying out my doctoral research, studying the devastating effects of social isolation on brain development in rodents. Being brought up alone prevented brain regions from connecting up fully and resulted in impaired memory, learning and decision-making whilst also inducing behaviors consistent with psychosis. It was upsetting and fascinating to observe that a lack of contact with others of their kind literally shrank the connectivity in their brains. Social connections and connections between brain cells were dependent on one another. We see something similar in humans - when we are part of an intelligent group, aspects our selfhood are able to reach their potential, and a sense of 'we' identity grows alongside our 'me' identity. It's a dance of mutual dependence; that which makes us entirely unique is nurtured by interactions with others, and in turn nurtures them.

We are all standing on the brink of an evolutionary step-change in human intelligence. The nature and timescale of that change is not yet determined. Technologists point to the 'singleton hypothesis' developed by philosopher Nick Bostrom of Oxford University, which predicts the emergence of a single world order. That might be a decision-making body such as a world government, a convergence of intelligent life forms through a shared moral code or a form of artificial intelligence. Whether the singleton would be hostile, neutral or benign towards humanity is, according to Bostrom, uncertain.

The singleton hypothesis is primarily a thought experiment but, as we've seen, there is emerging evidence from fields as diverse as neuroscience, epigenetics and economics that the evolution of human behaviour is speeding up and being shaped less by our genes and more by the influence of our environment. Increasingly it is what we learn from others and how we adapt to our transforming context that shapes the way humans think.

Tim Waring studies the evolution of strategies of cooperation and is based in the School of Economics at the University of Maine. He was one of the co-authors of the groundbreaking study into the shift away from genes and towards culture as the driving force in our species' development. He and his co-author, Zachary Wood, concluded in a paper published by the Royal Society in 2021 that as culture takes the lead, human beings' group identity and collective behaviour will become more crucial than our individual selfhood. 'In the very long term,' their report concluded, 'we suggest that humans are evolving from individual genetic organisms to cultural groups which function as super-organisms, similar to ant colonies and beehives.'

That future is a very long way off but the direction of travel is becoming clear. The skills that we already need, and that future generations will increasingly depend on, revolve around our flexibility of mind, our capacity to take on board new information and different points of view and our ability to collaborate. In an age of acute environmental pressures, the sustainability of human culture requires a shift away from 'me' to 'we' thinking. The world we live in has changed and our thinking must change to match. Only in this way will we be able to participate in what Waring calls the 'gene-culture co-evolution' of human intelligence.

Humanity has evolved for 'we' thinking. In recent times it's been harder to practise but we still instinctively know its power, embedded in the wisdom of our ancestors and passed on in parables like the one about a farmer who grew prize-winning corn. Every year, for twenty years, this farmer won an award at the local fair. One year a reporter interviewed him and learned something interesting about how he grew his excellent crop. The reporter discovered that the farmer shared his seed corn with his neighbours. 'How can you afford to share your best seed corn when your neighbours are entering corn in competition with yours each year?' she asked. 'The thing is,' replied the farmer, 'the wind picks up pollen from the ripening corn and swirls it from field to field. If my neighbours grow inferior corn, cross-pollination will steadily degrade the quality of mine. If I am to grow good corn, I must help my neighbours grow good corn.'

I love a good parable, always have. As a child I felt comforted by their compassionate wisdom, passed on from previous generations and from across the world. As an adult I love them even more, especially when they tie in with discoveries in neuroscience, as they nearly always do. So, just as cross-pollination means that no crop can be cultivated in isolation, the mechanism of emotional contagion shows us that feelings and thoughts, even moral codes, can be transmitted across neighbourhood fences and through friendships. Connectomics and genomics have shown us that we need the swirling wind of random genetic changes to ensure cognitive diversity is not stifled. Otherwise we will end up with a clone-like population whose homogenous way of thinking will make it more vulnerable to manipulation and less resistant to new threats. When the ancient wisdom that we intuitively know to be true is backed up by science, that feels reassuring to me. I can picture how these different forms of knowledge mesh in my mind, and the visualisation of the mechanisms that underpin the process helps to cement the ideas they contain.

Collective intelligence flourishes when we comprehend how interconnected we are, not only with others of our own species but with the whole natural world, and what amazing cognitive powers we can draw on when we join our thinking together. Every one of us can practise the skills we need to ensure our brains are fit for the future. Every one of us can play a small part in driving the evolution – and the return – from 'me' to 'we' thinking. The Earth *is* pivoting below our feet: it's time for the illusion of the importance of self to enter its final sunset.

Acknowledgements

Huge thanks go to Noosa, Australia for providing an unexpected place to write this book – a patch of paradise during the uncertainty of the pandemic. Thoughts are shaped by context and the people that surround us, and our adopted family there (Ros and John, Julia, Mackenzie, Laughlin, Baz, Tannelle, Nan Nan and Ross), and the friends we made along the way (Craig, Roxy, Bau, Georgia, Laura, Mia, Will, Alysha, Ella, Connor, Melanie, Eleanor, Naomi, Michael and Stella) helped to shape this book with their laughter, love and adventures. The Noosaville Library staff were incredibly helpful in accessing publications for research, whilst the rich Australian landscape, with its extreme climate, provided inspiration. It's a place where survival has depended on wisdom being passed across generations and between different groups of people. This is encapsulated in 'The Place', a dot painting by Alicia Adams, a Kamilaroi woman from the East coast of Australia whose artwork appears on the dedication page, is discussed in the Prologue, and nodded to in the image on the book's cover.

The field of research into collective intelligence has arisen from decades of study by scientists working across the globe, contributing millions of hours to produce a type of cathedral thinking that continues to develop. This book highlights key themes and take-home points from this body of knowledge. Specific studies feature as examples and huge thanks must go to the researchers who conducted this work, many of whom also spent time patiently and enthusiastically answering my questions and reading through relevant draft sections of the manuscript. They include: Barbara Sahakian, Robin Dunbar, Nichola Raihani, David Simons and Chris Chobris, Laura Cirelli, Vicky Leong, Sarah Jayne Blakemore, Alison Gopnik, Hao Zhang, Miguringa Sur, Paula Raymond and Daniel Wegner, Chad Sparber, Brain Uzi, Jane Goodall, Claire Stevenson, Juliet Harris, Holger Patzelt, Chris Frith, Taha Yasseri, Jens Krause, Adam Galinsky, Cameron Anderson, Jean Deceety, Deborah Gruenfeld, Anita Woolley, Molly Crockett, Ray Dolan, Marcela Litcanu, Layla Mofrad, Sigal Barsade, Adam Grant, Francisco Gino, Carol Dweck, Uffe Schjoedt and Andreas Roepstorff, William von Hippel, Sarah Garfinkle, Joel Pearson, Jennifer Eberhardt, John Coates, Jack Soll, Framingham Heart Study, Tanja Wingenbach, Fieldman Barrett and Crivelli's, Dacher Keltner, Emilie Caspar, Aline Vater, Julian Savulescu, Tad Oreszczyn, Agne kajackaite and Tom Chang, Dominik Mischkowski, Tomas Brodin, Amy Obden, Per Block and Stephanie Burnett Heyes, Victoria Spring, Sander van der Linden, Max Rollwage, Julia Wilson, SJ Beard, Robert Anda, Kerry Ressler, Rachel Yehuda, Karin Roelofs, Bridget Callaghan, Tim Bredy, Joel Pearson, Miguel Nicolelis, Chantel Prat. Huge thanks also go to Dr Dervila Glynn and Dr Emma Yhnell for reading through this entire manuscript prior to publication and providing their thoughtful and insightful feedback. As ever, thanks go to all at Magdalene College, University of Cambridge for their continued support and companionship.

There are many more co-curators to *Joined-Up Thinking* – most importantly from Hodder, the editors (Helen Coyle and Rowena Webb) who were initially confronted with an unwieldy manuscript double the stipulated length. Helen's pinpoint precision with emotionally intelligent highlighting, editing, and sculpting, alongside Rowena's encouragement and wisdom, honed this book into one we are all now exceptionally proud of. Thanks also go to Caroline Michel, literary agent extraordinaire, for her continued support and insight. Lastly, I'd like to thank Captain Mark Nash for his continued friendship and helping to keep life on an even, happy, keel.

References

Works are referred to in the order in which citations appear in the text.

PROLOGUE

The Flynn Effect and its recent reversal:

Baker, David P. *et al.* (2015) 'The cognitive impact of the education revolution: A possible cause of the Flynn Effect on population IQ.' *Intelligence*, 49: 144–58. doi:10.1016/j.intell.2015.01.003. ISSN 0160-2896

Flynn, J. R. (1984) 'The mean IQ of Americans: Massive rains 1932 to 1978.' *Psychological Bulletin*, 95, 29

Flynn, J. (1987) 'Massive IQ gains in 14 nations: What IQ tests really measure.' *Psychological Bulletin*, 101, 171

Lynn, R. (2009) 'Fluid intelligence but not vocabulary has increased in Britain.' *Intelligence*, 37, 249–55

Lynn, R. and Meisenberg, G. (2010) 'National IQs calculated and validated for 108 nations.' *Intelligence*, 38, 353–60

Te Nijenhuis, Jan and van der Flier, Henk. (2013) 'Is the Flynn effect on *g*?: A meta-analysis.' *Intelligence*, 41 (6), 802–7

UNESCO (2002) *Education for All: Is the World on Track?* https://unesdoc.unesco.org/ark:/48223/pf0000129053

Dutton, E., van der Linden, D., Lynn, R. (2016) 'The negative Flynn effect: A systematic literature review.' *Intelligence*, 59: 163–9

Bratsberg, Bernt and Rogeberg, Ole. 'Flynn effect and its reversal are both environmentally caused.' *PNAS*, 26 June 2018 115 (26) 6674–6678; first published 11 June 2018; https://doi.org/10.1073/pnas.1718793115

Moore, Oliver, 'Dumb and dumber: why we're getting less intelligent.' *The Times*, 12 June 2018, https://www.thetimes.co.uk/edition/news/dumb-and-dumber-why-we-re-getting-less-intelligent-80k3bl83v

Cognitive enhancement: our obsession with our own intelligence:

Brühl, A. B., Sahakian, B. J. (2016) 'Drugs, games, and devices for enhancing cognition: implications for work and society.' *Ann NY Acad Sci*; 1369(1):195–217. doi:10.1111/nyas.13040. Epub 4 April 2016. PMID: 27043232

Dresler, M. *et al.* (2019) 'Hacking the Brain: Dimensions of Cognitive Enhancement.' *ACS Chem Neurosci.*, 10(3):1137–48. doi:10.1021/acschemneuro.8b00571. Epub 2 January 2019. PMID: 30550256; PMCID: PMC6429408

Savulich, G. *et al.* (2017) 'Focusing the Neuroscience and Societal Implications of Cognitive Enhancers.' *Clin Pharmacol Ther*, 101(2):170–2. doi:10.1002/cpt.457. Epub 23 Sept 2016. PMID: 27557349

'Poll results: look who's doping'. Published online 9 April 2008. *Nature* 452, 674–5 (2008) | doi:10.1038/452674a, https://www.nature.com/news/2008/080409/full/452674a.html

Kodsi, Daniel. 'Revealed: Oxford's addiction to study drugs: 15 per cent of students have knowingly taken a "study drug", according to a Cherwell recent survey.' 13 May 2016. https://cherwell.org/2016/05/13/revealed-oxfords-addiction-to-study-drugs

Sahakian, Barbara. 'Opinion: Fair play? How "smart drugs" are making workplaces more competitive. 6 July 2016, https://www.cam.ac.uk/research/news/opinion-fair-play-how-smart-drugs-are-making-workplaces-more-competitive

Molteni, Megan. 'Netflix's "Unnatural Selection" Trailer Makes Crispr Personal: A new docuseries digs into the existential promise and peril of the gene-editing revolution.' *Science*, 10 April 2019. https://www.wired.com/story/netflixs-unnatural-selection-trailer-makes-crispr-personal

Regalado, Antonio. 'Chinese scientists are creating CRISPR babies.' *MIT Technology Review*, 25 November 2018

Bulluck, Pam. 'Gene-Edited Babies: What a Chinese Scientist Told an American Mentor.' *New York Times*, 14 April 2019. Retrieved 14 April 2019
https://www.theguardian.com/science/2021/oct/17/polygenic-screening-of-embryos-is-here-but-is-it-ethical

Zhou, M. *et al.* (2016) 'CCR5 is a suppressor for cortical plasticity and hippocampal learning and memory.' 5:e20985. doi:10.7554/eLife.20985

Lei, Shi *et al.* (2019) 'Transgenic rhesus monkeys carrying the human *MCPH1* gene copies show human-like neoteny of brain development.' *National Science Review*, vol. 6, issue 3, 480–93. https://doi.org/10.1093/nsr/nwz043

Wilson, Clare. (2018) 'Exclusive: A new test can predict IVF embryos' risk of having a low IQ: A new genetic test that enables people having IVF to screen out embryos likely to have a low IQ or high disease risk could soon become available in the US.' https://www.newscientist.com/article/mg24032041-900-exclusive-a-new-test-can-predict-ivf-embryos-risk-of-having-a-low-iq
https://www.technologyreview.com/2019/11/08/132018/polygenic-score-ivf-embryo-dna-tests-genomic-prediction-gattaca

Whitehouse, Andrew 'Prenatal screening and autism.' 17 November 2013
https://theconversation.com/prenatal-screening-and-autism-20395.

Ne'eman, Ari. 'Screening sperm donors for autism? As an autistic person, I know that's the road to eugenics.' *Guardian*, 30 December 2015. https://www.theguardian.com/commentisfree/2015/dec/30/screening-sperm-donors-autism-autistic-eugenics

Scangos, K. W. *et al.* (2021) 'Closed-loop neuromodulation in an individual with treatment-resistant depression.' *Nat Med* 27, 1696–1700. https://doi.org/10.1038/s41591-021-01480-w

Calyx, Cobi. (2020) 'Sustaining Citizen Science beyond an Emergency.' *Sustainability* 12, 4522; doi:10.3390/su12114522

Strasser, B. and Haklay, M. E. (2018) 'Citizen Science: Expertise, Democracy, and Public Participation.' SSC Policy Analysis 1/2018, 1–92

1. The Power of Joined-up Thinking

Neuroscience Is Only Recently Starting to Look at Brains Working Together

New Open Access journal in the field of Collective Intelligence, 4 August 2020. https://www.nesta.org.uk/press-release/sage-and-association-computing-machinery-announce-new-open-access-journal-field-collective-intelligence-collaboration-nesta/?gclid=Cjo KCQiAxc6PBhCEARIsA foT7Wvzw9Ra6khmJ5Yb9HwoDBCH chdet88fYme93hARZMKx62LsBdoaArrpEALw_wcB

Critchlow, Hannah (2018) *Consciousness: A Ladybird Expert Book* (The Ladybird Expert Series 29). London, UK: Michael Joseph

Brain Synchronisation

Denworth, Lydia (2019) '"Hyperscans" Show How Brains Sync as People Interact: Social neuroscientists ask what happens at the level of neurons when you tell someone a story or a group watches movies.' https://www.scientificamerican.com/article/hyperscans-show-how-brains-sync-as-people-interact

Montague, P. Read *et al.* (2002) 'Hyperscanning: simultaneous fMRI during linked social interactions.' *NeuroImage*, 16 (4): 1159–1164. doi:10.1006/nimg.2002.1150. ISSN 1053-8119. PMID: 12202103. S2CID: 15988039

Hasson, Uri *et al.* (2012) 'Brain-to-brain coupling: a mechanism for creating and sharing a social world.' *Trends in Cognitive Sciences*, 16 (2): 114–121. doi:10.1016/j.tics.2011.12.007. ISSN 1364-6613. PMC 3269540. PMID:22221820

Hu, Yi *et al.* (2018) 'Inter-brain synchrony and cooperation context in interactive decision making.' *Biological Psychology*, 133: 54–62. doi:10.1016/j.biopsycho.2017.12.005. ISSN 1873-6246. PMID:2929 2232. S2CID: 46859640

Liu, Difei *et al.* (2018) 'Interactive Brain Activity: Review and Progress on EEG-Based Hyperscanning in Social Interactions.' *Frontiers in Psychology*, 9: 1862. doi:10.3389/fpsyg.2018.01862. ISSN 1664-1078. PMC 6186988. PMID:30349495

Leong, Victoria *et al.* (2017) 'Speaker gaze increases information coupling between infant and adult brains.' *PNAS* 114 (50) 13290–13295; first published 28 November 2017; https://doi.org/10.1073/pnas.1702493114

Davidesco, Ido *et al.* 'Brain-to-brain synchrony between students and teachers predicts learning outcomes.' bioRxiv 644047; doi:https://doi.org/10.1101/644047

Davidesco, Ido *et al.* 'Brain-to-brain synchrony predicts long-term memory retention more accurately than individual brain measures.' doi:https://doi.org/10.1101/644047

Valencia, Ana Lucía, Froese, Tom. (2020) 'What binds us? Inter-brain neural synchronization and its implications for theories of human consciousness.' *Neuroscience of Consciousness*, vol. 2020, issue 1, niaao10, https://doi.org/10.1093/nc/niaao10

Hirsch, Joy *et al.* (2021) 'Interpersonal Agreement and Disagreement During Face-to-Face Dialogue: An fNIRS Investigation.' *Frontiers in Human Neuroscience*,14, https://www.frontiersin.org/article/10.3389/fnhum.2020.606397, doi:10.3389/fnhum.2020.606397

Cutting-edge Neuroscience Towards Collective Intelligence

Yu, H. *et al.* (2014) 'The voice of conscience: Neural bases of interpersonal guilt and compensation.' *Social Cognitive and Affective Neuroscience*, 9(8), 1150–8. (journal link)

Yu, H. *et al.* (2020) 'A generalizable multivariate brain pattern for interpersonal guilt.' *Cerebral Cortex*, 30(6), 3558-3572. (journal link) (pre-print)

Nicolle, A. *et al.* (2011) 'A role for the striatum in regret-related choice repetition.' *J Cogn Neurosci*.;23(4):845–56.doi:10.1162/jocn.2010.21510

Lufityanto, Galang, Donkin, Chris, Pearson, Joel. (2016) 'Measuring Intuition: Nonconscious Emotional Information Boosts Decision Accuracy and Confidence.' *Psychol Sci*, 27(5):622–34. doi:10.1177/0956797616629403. Epub 6 April 2016

Suchiya, Naotsugu; Koch, Christof (2004) "Continuous flash suppression." Vision Sciences Society Annual Meeting Abstract.' *Journal of Vision*, vol. 4, 61. doi:https://doi.org/10.1167/4.8.61

Vlassova, Alexandra, Donkin, Chris, Pearson, Joel. (2014) 'Unconscious information changes decision accuracy but not confidence.' *Proc*

Natl Acad Sci USA 11 Nov;111(45):16214–8. doi:10.1073/pnas.1403619111. Epub 27 Oct 2014

Quadt, L. *et al.* (2021) 'Interoceptive training to target anxiety in autistic adults (ADIE): A single-center, superiority randomized controlled trial.' *EClinicalMedicine*, *39*, 101042. https://doi.org/10.1016/j.eclinm.2021.101042

Organic Brain Integration With AI

Grau, Carles *et al.* (2014) 'Conscious Brain-to-Brain Communication in Humans Using Non-Invasive Technologies.' *PLoS One*, 9 (8): e105225 doi:10.1371/journal.pone.0105225

Renton, Angela I., Mattingley, Jason B. and Painter, David R. (2019). 'Optimising non-invasive brain-computer interface systems for free communication between naïve human participants.' *Scientific Reports*, 9 (1) 18705, 18705. doi:10.1038/s41598-019-55166-y

Jiang, L. *et al.* (2019) 'BrainNet: A Multi-Person Brain-to-Brain Interface for Direct Collaboration Between Brains.' *Sci Rep* 9, 6115. https://doi.org/10.1038/s41598-019-41895-7

Rao, R. P. *et al.* (2014) 'A direct brain-to-brain interface in humans.' *PLoS One* 9:e111332. 10.1371/journal.pone.0111332

Stocco, A. *et al.* (2015) 'Playing 20 questions with the mind: collaborative problem solving by humans using a brain-to-brain interface.' *PLoS One* 10:e0137303. 10.1371/journal.pone.0137303

Pais-Vieira, M. *et al.* (2013) 'Brain-to-Brain Interface for Real-Time Sharing of Sensorimotor Information.' *Scientific Reports* 3:, 1319

Pais-Vieira, M. *et al.* (2015) 'Building an organic computing device with multiple interconnected brains.' *Sci Rep.* 9 Jul;5:11869. doi:10.1038/srep11869. Erratum in: *Sci Rep.* 2015;5:14937. PMID: 26158615; PMCID: PMC4497302

Ramakrishnan, A. *et al.* (2015) 'Computing Arm Movements with a Monkey Brainet.' *Sci Rep* 5, 10767 https://doi.org/10.1038/srep10767

O'Doherty, J.E. *et al.* (2011) 'Active tactile exploration using a brain–machine–brain interface.' *Nature* 479: 228–31

Deadwyler, S. A. *et al.* (2013) 'Donor/recipient enhancement of memory in rat hippocampus.' *Front Syst Neurosci* 7: 120

Arjun Ramakrishnan et al. (2015) 'Computing Arm Movements with a Monkey Brainet'. Scientific Reports 5, article number: 10767; doi: 10.1038/srep10767

Miguel Pais-Vieira et al. (2015) 'Building an organic computing device with multiple interconnected brains'. Scientific Reports 5, article number: 11869; doi: 10.1038/srep11869

Jiang, L. et al. (2019) 'BrainNet: A Multi-Person Brain-to-Brain Interface for Direct Collaboration Between Brains.' Sci Rep 9, 6115. https://doi.org/10.1038/s41598-019-41895-7

The Effects of Social Isolation on Brain and Body

Fiorenzato, Eleonora et al. 'Impact of COVID-19-lockdown and vulnerability factors on cognitive functioning and mental health in Italian population:' doi:https://doi.org/10.1101/2020.10.02.20205237 Pre-print.

RESONANCE Consortium. 'Impact of the COVID-19 Pandemic on Early Child Cognitive Development: Initial Findings in a Longitudinal Observational Study of Child Health.' medRxiv. doi:https://doi.org/10.1101/2021.08.10.21261846; this version posted 11 August 2021.

Ingram, Joanne, Hand, Christopher J., Maciejewski, Greg. (2021) 'Social isolation during COVID-19 lockdown impairs cognitive function.' Journal of Applied Cognitive Psychology. https://doi.org/10.1002/acp.3821 https://onlinelibrary.wiley.com/doi/10.1002/acp.3821

Orben, A., Tomova, L., Blakemore, S. J. (2020) 'The effects of social deprivation on adolescent development and mental health.' Lancet Child Adolesc Health. 4(8):634–640. doi:10.1016/S2352-4642(20)30186-3. Epub 12 June 2020. PMID: 32540024; PMCID: PMC7292584

Zhang, S. X. et al. (2020) 'Unprecedented disruption of lives and work: Health, distress and life satisfaction of working adults in China one month into the COVID-19 outbreak.' Psychiatry Research, 288, 112958. https://doi.org/10.1016/j.psychres.2020.112958

Zunzunegui, M.-V. et al. (2003) 'Social networks, social integration, and social engagement determine cognitive decline in community-dwelling

Spanish older adults.' *Journal of Gerontology*, 58B, S93–S100. https://doi. org/10.1093/geronb/58.2.s93

Cacioppo, J. T. *et al.* (2000) 'Lonely traits and concomitant physiological processes: The MacArthur social neuroscience studies.' *International Journal of Physiology*, 35(2–3), 143–54. https://doi.org/10.1016/s0167-8760(99)00049-5

Evans, I. E. M. *et al.* (2018) 'Social isolation, cognitive reserve, and cognition in healthy older people.' *PLoS One*, 13(8), e0201008. http://dx.doi.org/10.1371/journal.pone.0201008

A landmark study in the US tracked individuals and found those reported being lovely more likely to be depressed five years later: Cacioppo J. T., Hawkley L. C., Thisted, R. A. (2010) 'Perceived social isolation makes me sad: 5-year cross-lagged analyses of loneliness and depressive symptomatology in the Chicago Health, Aging, and Social Relations Study.' *Psychol Aging*. 25(2):453–63. doi:10.1037/a0017216. PMID: 20545429; PMCID: PMC2922929

Meltzer, H. *et al.* (2013) 'Feelings of loneliness among adults with mental disorder.' *Soc Psychiatry Psychiatr Epidemiol*. 48(1):5-13. doi:10.1007/s00127-012-0515-8. Epub 9 May 2012. Erratum in: S*oc Psychiatry Psychiatr Epidemiol*. 2015 Mar;50(3):503-4. PMID:22570258

Silk, J. B. 'Evolutionary Perspectives on the Links Between Close Social Bonds, Health, and Fitness.' In: Committee on Population; Division of Behavioral and Social Sciences and Education; National Research Council; Weinstein M., Lane M.A. (eds). *Sociality, Hierarchy, Health: Comparative Biodemography: A Collection of Papers*. Washington, D. C.: National Academies Press (US); 22 September 2014. 6. Available from: https://www.ncbi.nlm.nih.gov/books/NBK242452

Johnson, Zachary and Young, Larry. (2015) 'Neurobiological mechanisms of social attachment and pair bonding.' *Current Opinion in Behavioral Sciences*. 3. 38–44. 10.1016/j.cobeha.2015.01.009

Cacioppo, J. T. *et al.* (2009) 'In the eye of the beholder: individual differences in perceived social isolation predict regional brain activation to social stimuli.' *J Cogn Neurosci*. 21(1):83–92. doi:10.1162/jocn.2009.21007. PMID: 18476760; PMCID: PMC2810252

Cacioppo, J. T., Chen, H. Y., Cacioppo, S. (2017) 'Reciprocal Influences Between Loneliness and Self-Centeredness: A Cross-Lagged Panel

Analysis in a Population-Based Sample of African American, Hispanic, and Caucasian Adults.' *Pers Soc Psychol Bull.* 43(8):1125–1135. doi:10.1177/0146167217705120. Epub 2017 Jun 13. PMID: 28903715

The Social Brain and Collective Intelligence Underpinning the Next Stage of Our Species Evolution

Frith, U., Frith, C. (2010) 'The social brain: allowing humans to boldly go where no other species has been.' *Philos Trans R Soc Lond B Biol Sci.* 365(1537):165–76. doi:10.1098/rstb.2009.0160

Waring, Timothy M. and Wood, Zachary T. (2021) 'Long-term gene–culture coevolution and the human evolutionary transition.' *Proc. R. Soc. B.* 2882021053820210538, http://doi.org/10.1098/rspb.2021.0538

Andersson, C., Törnberg, P. (2008) 'Toward a macroevolutionary theory of human evolution: the social protocell.' *Biol. Theory* 14, 86–102. doi:10.1007/s13752-018-0313-y

Gowdy, J., Krall, L. (2014) 'Agriculture as a major evolutionary transition to human ultrasociality.' *J. Bioecon.* 16, 179–202. doi:10.1007/s10818-013-9156-6

Maynard Smith, J., Szathmáry, E. (1995) *The Major Transitions in Evolution.* Oxford, UK: W. H. Freeman Spektrum

Powers, S. T., van Schaik, C. P., Lehmann, L. (2016) 'How institutions shaped the last major evolutionary transition to large-scale human societies.' *Phil Trans R Soc B.* 371, 20150098. doi:10.1098/rstb.2015.0098

Stearns, S. C. (2007) 'Are we stalled part way through a major evolutionary transition from individual to group?' *Evolution* 61, 2275–80. doi:10.1111/j.1558-5646.2007.00202.x

Szathmáry, E. (2015) 'Toward major evolutionary transitions theory 2.0.' *Proc Natl Acad Sci USA* 112, 10 104–10 111. doi:10.1073/pnas.1421398112

Calcott, B., Sterelny, K. (eds). (2011) *The major Transitions in Evolution Revisited*, 1st edn. Cambridge, MA: MIT Press. See https://ebookcentral.proquest.com/lib/umaine/reader.action?docID=3339240&ppg=180

Michod, R. E. (2000) *Darwinian Dynamics: Evolutionary Transitions in Fitness and Individuality.* Princeton, NJ: Princeton University Press.

Queller, D. C., Strassmann J. E. (2009) 'Beyond society: the evolution of organismality.' *Phil. Trans. R. Soc. B* 364, 3143–55. doi:10.1098/rstb.2009.0095

West, S. A., Fisher, R. M., Gardner, A., Kiers E. T. (2015) 'Major evolutionary transitions in individuality.' *Proc Natl Acad Sci USA* 112, 10 112–10 119. doi:10.1073/pnas.1421402112

Okasha, S. (2005) 'Multilevel selection and the major transitions in evolution.' *Philos. Sci.* 72, 1013–25. doi:10.1086/508102

Kesebir, S. (2012) 'The superorganism account of human sociality: how and when human groups are like beehives.' *Pers Soc Psychol Rev.* 16, 233–61. doi:10.1177/1088868311430834

All Brains Are Not Alike

Gopnik, A. *et al.* (2017) 'Changes in cognitive flexibility and hypothesis search across human life history from childhood to adolescence to adulthood.' *Proc Natl Acad Sci USA*;114(30):7892–9. doi:10.1073/pnas.1700811114

Blakemore, Sarah-Jayne (2018) *Inventing Ourselves: The Secret Life of the Teenage Brain* (first edition). Doubleday

Critchlow, Hannah (2019) *The Science of Fate: The New Science of Who We Are – And How to Shape our Best Future.* London, UK: Hodder & Stoughton

Kempermann, G. (2019) 'Environmental enrichment, new neurons and the neurobiology of individuality.' *Nat Rev Neurosci.* 20(4):235–45. doi:10.1038/s41583-019-0120-x. PMID: 30723309

Becht, A. I., Mills, K.L. (2020) 'Modeling Individual Differences in Brain Development.' *Biol Psychiatry.* 88(1):63–9. doi:10.1016/j.biopsych.2020.01.027. Epub 11 Feb 2020. PMID: 32245576; PMCID: PMC7305975

Critchlow, Hannah (2018) *Consciousness: A Ladybird Expert Book* (The Ladybird Expert Series 29). London, UK: Michael Joseph

Ciarrusta, J., Dimitrova, R., Batalle, D. et al. (2020) Emerging functional connectivity differences in newborn infants vulnerable to autism spectrum disorders. Transl Psychiatry 10, 131. https://doi.org/10.1038/s41398-020-0805-y

Frank, M. J. *et al.* (2009) 'Prefrontal and striatal dopaminergic genes predict individual differences in exploration and exploitation.' *Nat. Neurosci.* 12, 1062–8. doi:10.1038/nn.2342

Badre, D. *et al.* (2012) 'Rostrolateral prefrontal cortex and individual differences in uncertainty-driven exploration.' *Neuron* 73, 595–607. doi:10.1016/j.neuron.2011.12.025

Indigenous artwork

Adams, Alicia NAIDOC Exhibition First Nations Artists connected to Kabi Kabi/Gubbi Gubbi country exhibition at the Cooroy Butter Factory Arts Centre, QLD, Australia 18 June–18 July 2021, Heal Country: http://www.butterfactoryartscentre.com.au/exhibition-archive.html

2. What Is This Thing Called Intelligence?

All Brains Are Not Alike

Ritchie, Stuart (2015) *Intelligence: All That Matters* (first edition). London, UK: Hodder & Stoughton

https://www.ninds.nih.gov/Disorders/Patient-Caregiver-Education/Genes-Work-Brain

Kempermann, G. (2019) 'Environmental enrichment, new neurons and the neurobiology of individuality.' *Nat Rev Neurosci.* 20(4):235–45. doi:10.1038/s41583-019-0120-x. PMID: 30723309

Becht, A. I., Mill,s K.L. (2020) 'Modeling Individual Differences in Brain Development.' *Biol Psychiatry.* 1;88(1):63–9. doi:10.1016/j.biopsych.2020.01.027. Epub 11 Feb 2020. PMID: 32245576; PMCID: PMC7305975

Critchlow, Hannah (2018) *Consciousness: A Ladybird Expert Book* (The Ladybird Expert Series 29). London, UK: Michael Joseph

Ciarrusta, J., Dimitrova, R., Batalle, D. et al. (2020) Emerging functional connectivity differences in newborn infants vulnerable to autism spectrum disorders. Transl Psychiatry 10, 131. https://doi.org/10.1038/s41398-020-0805-y

Baker, J. T. *et al.* (2019) 'Functional connectomics of affective and psychotic pathology.' *Proc Natl Acad Sci USA.* 116(18):9050–9. doi:10.1073/pnas.1820780116. Epub 15 April 2019. PMID: 30988201; PMCID: PMC6500110

Frank, M. J. *et al.* (2009) 'Prefrontal and striatal dopaminergic genes predict individual differences in exploration and exploitation.' *Nat. Neurosci.* 12, 1062–8. doi:10.1038/nn.2342

Badre, D. *et al.* (2012) 'Rostrolateral prefrontal cortex and individual differences in uncertainty-driven exploration.' *Neuron* 73, 595–607. doi:10.1016/j.neuron.2011.12.025

Sedgwick, J. A., Merwood, A. and Asherson, P. (2019) 'The positive aspects of attention deficit hyperactivity disorder: a qualitative investigation of successful adults with ADHD.' *ADHD Atten Def Hyp Disord* 11, 241–53. https://doi.org/10.1007/s12402-018-0277-6

White, H. A., Shah, P. (2006) 'Uninhibited imaginations: creativity in adults with attention-deficit/hyperactivity disorder.' *Pers Individ Differ* 40:1121–31

Brain Changes Over the Course of the Lifespan

Gopnik, A. *et al.* (2017) 'Changes in cognitive flexibility and hypothesis search across human life history from childhood to adolescence to adulthood.' *Proc Natl Acad Sci USA.* 114(30):7892–9. doi:10.1073/pnas.1700811114

Blakemore, Sarah-Jayne (2018) *Inventing Ourselves: The Secret Life of the Teenage Brain* (first edition). Doubleday

Critchlow, Hannah (2019) *The Science of Fate: The New Science of Who We Are – And How to Shape our Best Future*. London, UK: Hodder & Stoughton.

Beyond reason – For a Diversity of Intelligences

Wilson, Siân *et al.* (2021) 'Development of human white matter pathways in utero over the second and third trimester.' *PNAS*

Eyre, M. *et al.* (2021) 'The Developing Human Connectomme Project: typical and disrupted perinatal functional connectivity.' *Brain*

Ciarrusta, J., Dimitrova, R., Batalle, D. et al. (2020) Emerging functional connectivity differences in newborn infants vulnerable to autism spectrum disorders. Transl Psychiatry 10, 131. https://doi.org/10.1038/s41398-020-0805-y

Holtmaat, A., Svoboda, K. (2009) 'Experience-dependent structural synaptic plasticity in the mammalian brain.' *Nat Rev Neurosci*;10:647–58

REFERENCES

Matsuzaki, M. *et al.* (2004) 'Structural basis of long-term potentiation in single dendritic spines.' *Nature.* 429:761–6

Dromard, Y. *et al.* (2021) 'Dual imaging of dendritic spines and mitochondria *in vivo* reveals hotspots of plasticity and metabolic adaptation to stress.' *Neurobiol Stress.* 15:100402. doi:10.1016/j.ynstr.2021.100402. PMID: 34611532; PMCID: PMC8477201

Sadakane, O. *et al.* (2015) 'In Vivo Two-Photon Imaging of Dendritic Spines in Marmoset Neocortex.' *eNeuro*, 2(4):ENEURO.0019-15.2015. doi:10.1523/ENEURO.0019-15.2015

Mizrahi, A. *et al.* (2004) 'High-resolution in vivo imaging of hippocampal dendrites and spines.' *J Neurosci.* 24(13):3147–51. doi:10.1523/JNEUROSCI.5218-03.2004

Gu, L. *et al.* (2014) 'Long-term in vivo imaging of dendritic spines in the hippocampus reveals structural plasticity.' *J Neurosci.* 34(42):1394853. doi:10.1523/JNEUROSCI.1464-14.2014

Yang, G., Pan F., Gan, W. B. (2009) 'Stably maintained dendritic spines are associated with lifelong memories.' *Nature.*;462:920–4. doi:10.1038/nature08577

Lai, C. S., Franke, T. F., Gan, W. B. (2012) 'Opposite effects of fear conditioning and extinction on dendritic spine remodelling.' *Nature.* 483:87–91. doi:10.1038/nature10792

Trachtenberg, J. T. *et al.* (2002) 'Long-term in vivo imaging of experience-dependent synaptic plasticity in adult cortex.' *Nature.* 420(6917):788–94. doi:10.1038/nature01273. PMID: 12490942

Fox, M. E. *et al.* (2020) 'Dendritic spine density is increased on nucleus accumbens D2 neurons after chronic social defeat.' *Sci Rep.* 10(1):12393. doi:10.1038/s41598-020-69339-7. PMID: 32709968; PMCID: PMC7381630

Critchlow, H. M. 'The Role of Dendritic Spine Plasticity in Schizophrenia' (doctoral thesis, University of Cambridge, 2007).

Markett, S. *et al.* (2020) 'Specific and segregated changes to the functional connectome evoked by the processing of emotional faces: A task-based connectome study.' *Sci Rep* 10, 4822 https://doi.org/10.1038/s41598-020-61522-0

Bennett, Sophie H., Kirby, Alastair J., Finnerty, Gerald T. (2018) 'Rewiring the connectome: Evidence and effects.' *Neuroscience &*

Biobehavioral Reviews, vol. 88, 51–62, ISSN 0149-7634, https://doi.org/10.1016/j.neubiorev.2018.03.001

Why Groups Are Smarter

11 billion bytes of data enter our senses every second and other brain facts: https://www.britannica.com/science/information-theory/Physiology

Ahmadpoor, Mohammad and Jones, Benjamin F. (2019) 'Decoding team and individual impact in science and invention.' *PNAS*, 116 (28) 13885–90; first published 24 June 2019; https://doi.org/10.1073/pnas.1812341116

Bahrami, B. *et al.* (2010) 'Optimally interacting minds.' *Science*, 329 (5995), 1081–5. doi:10.1126/science.1185718

Ariely, D. *et al.* (2000) 'The effects of averaging subjective probability estimates between and within judges.' *J Exp Psychol Appl.* 6, 130–47. doi:10.1037/1076-898X.6.2.130

Johnson, T. R., Budescu, D. V., Wallsten, T. S. (2001) 'Averaging probability judgments: Monte Carlo analyses of asymptotic diagnostic value.' *J. Behav. Decis. Making* 14, 123–40. doi:10.1002/bdm.369

Galton, F. (1907) 'Vox populi.' *Nature* 75, 450–1. doi:10.1038/075450a0

Migdał, P. *et al.* (2012) 'Information-sharing and aggregation models for interacting minds.' *J Math Psychol* 56, 417–26. doi:10.1016/j.jmp.2013.01.002

Krause, J., Ruxton, G. D., Krause, S. (2010) 'Swarm intelligence in animals and humans.' *Trends Ecol Evol.* 25, 28–34. doi:10.1016/j.tree.2009.06.016

Couzin, I. D. (2009) 'Collective cognition in animal groups.' *Trends Cogn. Sci.* 13, 36–43. doi:10.1016/j.tics.2008.10.002

Sterzer, P., Frith, C. and Petrovic, P. (2010) 'Believing is seeing: expectations alter visual awareness.' *Current Biology*, 20 (21), 1973. vol. 18, R697, 2008. doi:10.1016/j.cub.2010.10.036

Bang, D. and Frith, C. (2017) 'Making better decisions in groups.' Royal Society Open Science. doi:10.1098/rsos.170193

Bahrami, B. *et al.* (2010) 'Optimally interacting minds.' *Science*, 329 (5995), 1081–5. doi:10.1126/science.1185718

Bang, D. *et al.* (2014) 'Does interaction matter? Testing whether a confidence heuristic can replace interaction in collective decision-making.' *Conscious Cogn*, 26, 13–23. doi:10.1016/j.concog.2014.02.002

Bahrami, B. *et al.* (2012) 'Together, slowly but surely: the role of social interaction and feedback on the build-up of benefit in collective decision-making.' *J Exp Psychol Hum Percept Perform*, 38 (1), 3–8. doi:10.1037/a0025708

Survival Is All About Networks

The first mention of the sociome: Kamiyama, D. 'Bioprobes and Genetics Reveal the Signal Integration that Initiates Dendrites in a Neuron in Vivo.' (Doctoral dissertation. Tokyo University of Science, (2001)

Lee, S. H. *et al.* (2020) 'Emotional well-being and gut microbiome profiles by enterotype.' *Sci Rep* 10, 20736 https://doi.org/10.1038/s41598-020-77673-z

Interoception Summit 2016 participants. (2018) 'Interoception and Mental Health: A Roadmap.' *Biol Psychiatry Cogn Neurosci Neuroimaging.* 3(6):501–13. doi:10.1016/j.bpsc.2017.12.004. Epub 28 Dec 2017. PMID: 29884281; PMCID: PMC6054486

Lufityanto, Galang, Donkin, Chris, Pearson, Joel (2016) 'Measuring Intuition: Nonconscious Emotional Information Boosts Decision Accuracy and Confidence.' *Psychol Sci*,27(5):622–34. doi:10.1177/0956797616629403. Epub 6 April 2016

'"Continuous flash suppression." Vision Sciences Society Annual Meeting Abstract.' (2004) Tsuchiya, Naotsugu, Koch, Christof. *Journal of Vision*, vol. 4, 61. doi:https://doi.org/10.1167/4.8.61

Vlassova, Alexandra, Donkin, Chris, Pearson, Joel (2014) 'Unconscious information changes decision accuracy but not confidence.' *Proc Natl Acad Sci USA* 111(45):16214–8. doi:10.1073/pnas.1403619111. Epub 27 Oct 2014

Quadt, L. *et al.* (2021) 'Interoceptive training to target anxiety in autistic adults (ADIE): A single-center, superiority randomized controlled trial.' *EClinicalMedicine*, 39, 101042. https://doi.org/10.1016/j.eclinm.2021.101042

Kandasamy, Narayanan *et al.* (2016) 'Interoceptive Ability Predicts

Survival on a London Trading Floor.' *Scientific Reports*, 6: 32986. doi:10.1038/srep32986

Intelligence Is Always Collective

'Trees Have Their Own Songs: A new book by David George Haskell invites us to listen.' *The Atlantic*, 4 April 2017, Ed Yong, https://www.theatlantic.com/science/archive/2017/04/trees-have-their-own-songs/521742

Critchlow, Hannah (2018) *Consciousness: A Ladybird Expert Book* (The Ladybird Expert Series 29). London, UK: Michael Joseph

Powell, J. L. *et al.* (2010) 'Orbital prefrontal cortex volume correlates with social cognitive competence.' *Neuropsychologia.* 48(12):3554–62. doi:10.1016/j.neuropsychologia.2010.08.004. Epub 14 Aug 2010. PMID: 20713074

Lewis, P. A. (2011) 'Ventromedial prefrontal volume predicts understanding of others and social network size.' *Neuroimage.* 57(4):1624–9. doi:10.1016/j.neuroimage.2011.05.030. Epub 15 May 2011. PMID: 21616156

Humanity's Unique Contribution to Collective Intelligence

The optimal group size and friendship groups: Professor Robin Dunbar and the Social Brain Hypothesis: as discussed in Critchlow, Hannah (2019), *The Science of Fate: The New Science of Who We Are – And How to Shape our Best Future.* London, UK: Hodder & Stoughton

Raihani, Nichola (2021) *The Social Instinct: How Cooperation Shaped the World.* London, UK: Jonathan Cape

Dawkins, Richard (1976) *The Selfish Gene.* Oxford, UK: Oxford University Press

Gladwell, Malcolm (2002) *The Tipping Point: How Little Things Can Make a Big Difference.* London, UK: Abacus

The Age of the Extended Mind

Clark, Andy and Chalmers, David (1998) 'The Extended Mind.' *Analysis*, vol. 58, no. 1, 7–19, http://www.jstor.org/stable/3328150. Accessed 9 May 2022

3. Family: The Cradle of Joined-up Thinking

The Monkey Business Illusion: https://www.youtube.com/watch?v=IGQmdoK_ZfY

Simons, Daniel J.; Chabris, Christopher F. (1999) 'Gorillas in our midst: sustained inattentional blindness for dynamic events.' (PDF). *Perception*. 28 (9): 1059–74. CiteSeerX 10.1.1.65.8130. doi:10.1068/p2952. PMID:10694957

Liu, Han-Hui (2018) 'Age-Related Effects of Stimulus Type and Congruency on Inattentional Blindness'. *Front Psychol*. https://doi.org/10.3389/fpsyg.2018.00794

Graham, E. R. and Burke, D. M. (2011) 'Aging increases inattentional blindness to gorillas in our midst.' *Psychology and Aging* 26(1): 162–6, doi:https://doi.org/10.1037/a0020647

Stothart, Cary; Boot, Walter; Simons, Daniel (2015) 'Using Mechanical Turk to Assess the Effects of Age and Spatial Proximity on Inattentional Blindness.' *Collabra*. 1 (1): 2. doi:10.1525/collabra.26

Graham, E. R. and Burke, D. M. (2011) 'Aging Increases Inattentional Blindness to the Gorilla in our Midst.' *Psychology and Aging*. 26(1): 162–6. doi:10.1037/a0020647

Are We Born Clever – Or Do We Learn Cleverness?

Binet, Alfred; Simon, Th. (1916) *The development of intelligence in children: The Binet–Simon Scale*. Publications of the Training School at Vineland New Jersey Department of Research no. 11. E. S. Kite (trans.). Baltimore: Williams & Wilkins

Becker, K. A. (2003) 'History of the Stanford–Binet Intelligence scales: Content and psychometrics.' Stanford–Binet Intelligence Scales (fifth edition), Assessment Service Bulletin no. 1

Johnson, W., McGue, M. and Iacono, W. G. (2006) 'Genetic and environmental influences on academic achievement trajectories during adolescence.' *Dev Psychol*. 42, 513–42

Strenze, T. (2007) 'Intelligence and socioeconomic success: a meta-analytic review of longitudinal research.' *Intelligence* 35, 401–26 (1997) AND Gottfredson, L. 'Why *g* matters: the complexity of everyday life.' *Intelligence* 24, 79–132

Batty, G. D., Deary, I. J. and Gottfredson, L. S. (2007) 'Premorbid (early life) IQ and later mortality risk: systematic review.' *Ann Epidemiol.* 17, 278–88

Batty, G. D. *et al.* (2009) IQ in late adolescence/early adulthood and mortality by middle age: cohort study of one million Swedish men.' *Epidemiology* 20, 100–109

Deary, Ian J., Pattie, Alison, Starr, John M. (2013) 'The Stability of Intelligence From Age 11 to Age 90 Years: The Lothian Birth Cohort of 1921.' *Psychological Science*, vol. 24, issue 12, pp 2361–8

Ritchie, Stuart. (2015) *Intelligence: All That Matters* (first edition). London, UK: Hodder & Stoughton

https://theconversation.com/the-iq-test-wars-why-screening-for-intelligence-is-still-so-controversial-81428

BBC Four Eugenics: Science's Greatest Scandal: Journalist Angela Saini and disability rights activist Adam Pearson explore the shocking origins and legacy of eugenics in Britain and its continued influence today. https://www.bbc.co.uk/programmes/m0008zc7

Tydén, Mattias (2002). *Från politik till praktik : de svenska steriliseringslagarna 1935–1975.* Stockholm studies in history, 0491–0842 ; 63. Stockholm: Almqvist & Wiksell International. 69–70

'The Role of Genes in the Brain:' https://www.ninds.nih.gov/Disorders/Patient-Caregiver-Education/Genes-Work-Brain

Kempermann, G. (2019) 'Environmental enrichment, new neurons and the neurobiology of individuality.' *Nat Rev Neurosci.* 20(4):235–45. doi:10.1038/s41583-019-0120-x. PMID: 30723309

Becht, A. I., Mill,s K. L. (2020) 'Modeling Individual Differences in Brain Development.' *Biol Psychiatry.* 88(1):63–9. doi:10.1016/j.biopsych.2020.01.027. Epub 11 Feb 2020. PMID: 32245576; PMCID: PMC7305975

Critchlow, Hannah (2018) *Consciousness: A Ladybird Expert Book* (The Ladybird Expert Series 29). London, UK: Michael Joseph

Ciarrusta, J. *et al.* (2020) 'Emerging functional connectivity differences in newborn infants vulnerable to Autism Spectrum Disorders.' *Translational Psychiatry*

Baker, J. T. *et al.* (2019) 'Functional connectomics of affective and psychotic pathology.' *Proc Natl Acad Sci USA.* 116(18):9050–9.

doi:10.1073/pnas.1820780116. Epub 15 Apr 2019. PMID: 30988201; PMCID: PMC6500110

Thompson, P. M. *et al.* (2001) 'Genetic influences on brain structure.' *Nat. Neurosci.* 4, 1253–8

Giedd, J. N. *et al.* (1999) 'Brain development during childhood and adolescence: a longitudinal MRI study.' *Nat Neurosci.* 2, 861–3

Sowell, E. R. *et al.* (2003) 'Mapping cortical change across the human life span.' *Nat Neurosci.* 6, 309–15; 10.1038/nn1008

Toga, A. W., Thompson, P. M. and Sowell, E. R. (2006) 'Mapping brain maturation.' *Trends Neurosci.* 29, 148 –59; 10.1016/j.tins.2006.01.007

Born to Be Empathetic: Babies

Premack, D., Woodruff, G. (1978) 'Does the chimpanzee have a theory of mind?' *Behav. Brain Sci.* 1 515–26. 10.1017/s0140525x00076512

Wellman, H. M., Cross, D., Watson, J. (2001) 'Meta-analysis of theory-of-mind development: the truth about false belief.' *Child Dev.* 72 655–684. 10.1111/1467-8624.00304

Carpendale, J. I. M., Lewis, C. (2004) 'Constructing an understanding of mind: the development of children's social understanding within social interaction.' *Behav Brain Sci.* 27 79–151. 10.1017/s0140525 x04000032

Feldman, R. (2007) 'Parent-infant synchrony and the construction of shared timing; Physiological precursors, developmental outcomes, and risk conditions.' *J Child Psychol Psychiatry* 48 329–54. 10.1111/j.1469-7610.2006.01701.x

National Collaborating Centre for Mental Health (UK) (2015) 'Children's Attachment: Attachment in Children and Young People Who Are Adopted from Care, in Care or at High Risk of Going into Care.' London: National Institute for Health and Care Excellence (NICE); Nov. (NICE Guideline, No. 26.) 2, Introduction to children's attachment. Available from: https://www.ncbi.nlm.nih.gov/books/NBK356196

Kragness, H. E., Johnson, E. K. and Cirelli, L. K. (in press). 'The song, not the singer: Infants prefer to listen to familiar songs, regardless of singer identity.' *Developmental Science*

Cirelli, L. K. and Trehub, S. E. (2018) 'Infants help singers of familiar songs.' *Music & Science,* 1. doi:10.1177/2059204318761622

Cirelli, L. K. (2018) 'How interpersonal synchrony facilitates early prosocial behavior.' *Current Opinion in Psychology*, 20, 35–9

Cirelli, L. K. *et al.* (2017) 'Effects of interpersonal movement synchrony on infant helping behaviours: Is music necessary?' *Music Perception*, 34, 319–26

Leong, V. *et al.* 'Speaker gaze changes information coupling between infant and adult brains'. *Proceedings of the National Academy of Sciences* 114 (50), 13290–5

Wass, S. V. *et al.* (2020) 'Interpersonal neural entrainment during early social interaction. *Trends in Cognitive Sciences* 24 (4), 329–42

Reindl, V. *et al.* (2022) 'Multimodal hyperscanning reveals that synchrony of body and mind are distinct in mother-child dyads.' *NeuroImage* 251, 118982

Zhu, Y., Leong, V., Hou, Y., Zhang, D., Pan, Y., & Hu, Y. (2021). Instructor -learner neural synchronization during elaborated feedback predicts learning transfer. Journal of Educational Psychology. Advance online publication. https://doi.org/10.1037/edu0000707

Haresign, I. M. *et al.* (2022) 'Measuring the temporal dynamics of inter-personal neural entrainment in continuous child-adult EEG hyperscanning data.' *Developmental Cognitive Neuroscience* 54, 101093

Leong, V., Ham, G. X., Augustine, G. J. (2017) 'Using Optogenetic Dyadic Animal Models to Elucidate the Neural Basis for Human Parent–Infant Social Knowledge Transmission.' *Frontiers in Neural Circuits*, 101

Leong, V. *et al.* 'Social Interaction in Neuropsychiatry.' *Frontiers in Psychiatry* 12, 526

Piazza, E. A. *et al.* (2020) 'Infant and adult brains are coupled to the dynamics of natural communication.' *Psychological Science*, *31*, 6–17. https://www.princeton.edu/news/2020/01/09/baby-and-adult-brains-sync-during-play-finds-princeton-baby-lab

Honey, C. J. *et al.* (2012) 'Not lost in translation: Neural responses shared across languages.' *Journal of Neuroscience* 32(44):15277–83

Regev, M., Honey, U., Hasson, U. (2013) 'Modality-selective and modality-invariant neural responses to spoken and written narratives.' *Journal of Neuroscience* 33(40):15978–88

Hasson, Uri 'This is your brain on communication' https://www.ted.com/talks/uri_hasson_this_is_your_brain_on_communication

Critchlow, Hannah (2018) *Consciousness: A Ladybird Expert Book* (The Ladybird Expert Series 29), London, UK: Michael Joseph

Natural Creatives: Teens

Blakemore, Sarah-Jayne (2018) *Inventing Ourselves: The Secret Life of the Teenage Brain* (first edition). Doubleday

Critchlow, Hannah (2019) *The Science of Fate: The New Science of Who We Are – And How to Shape our Best Future.* London, UK: Hodder & Stoughton

Ritchie, Stuart (2015) *Intelligence: All That Matters* (first edition). London, UK: Hodder & Stoughton

Stevenson, Claire, E. *et al.* (2014) 'Training creative cognition: adolescence as a flexible period for improving creativity.' *Frontiers in Human Neuroscience*, 8, https://www.frontiersin.org/article/10.3389/fnhum.2014.00827, doi:10.3389/fnhum.2014.00827 ISSN=1662-5161

Kleibeuker, S. W., de Dreu, C. K., Crone, E. A. (2016) 'Creativity Development in Adolescence: Insight from Behavior, Brain, and Training Studies.' *New Dir Child Adolesc Dev.* Spring; (151):73–84. doi:10.1002/cad.20148. PMID: 26994726

Stevenson, Claire, (2020) 'Are adolescents more creative than adults?' https://bold.expert/are-adolescents-more-creative-than-adults

Kleibeuker, S.W., de Dreu, C. K., Crone, E.A. (2013) 'The development of creative cognition across adolescence: distinct trajectories for insight and divergent thinking.' *Dev Sci.* 16(1):2–12. doi:10.1111/j.1467-7687.2012.01176.x. Epub 8 Oct 2012. PMID: 23278922

Stevenson, C. E., Kleibeuker, S. W., de Dreu, C. K., Crone, E. A. (2014) 'Training creative cognition: adolescence as a flexible period for improving creativity.' *Front Hum Neurosci.* 8:827. doi:10.3389/fnhum.2014.00827. PMID: 25400565; PMCID: PMC4212808

Wu, Chi Hang *et al.* (2005) 'Age Differences in Creativity: Task Structure and Knowledge Base' *Creativity Research Journal*, 17:4, 321–6, doi:10.1207/s15326934crj1704_3

Gopnik, A. *et al.* (2017) 'Changes in cognitive flexibility and hypothesis search across human life history from childhood to adolescence to adulthood.' *Proc Natl Acad Sci USA.*;114(30):7892–9. doi:10.1073/pnas.1700811114

Zhou, Yanyun *et al.* (2017) 'The Impact of Bodily States on Divergent Thinking: Evidence for a Control-Depletion Account.' *Frontiers in Psychology.* 8. 1546. 10.3389/fpsyg.2017.01546

Zhang, Hao, Liu, Jia and Zhang, Qinglin. (2013) 'Neural representations for the generation of inventive conceptions inspired by adaptive feature optimization of biological species.' *Cortex; a journal devoted to the study of the nervous system and behavior.* 50. 10.1016/j.cortex.2013.01.015

Gao, Ying and Zhang, Hao. (2014) 'Unconscious processing modulates creative problem solving: Evidence from an electrophysiological study.' *Consciousness and Cognition.* 26. 64–73. 10.1016/j.concog.2014.03.001

Wisdom and Expertise: Older People

Ritchie, Stuart (2015) *Intelligence: All That Matters* (first edition). London, UK: Hodder & Stoughton

Maguire, E. A. *et al.* (2000) 'Navigation-related structural change in the hippocampi of taxi drivers.' *Proc Natl Acad Sci USA.* 97(8):4 398-403. doi:10.1073/pnas.070039597. PMID: 10716738; PMCID: PMC18253

El-Boustani S. *et al.* (2018) 'Locally coordinated synaptic plasticity of visual cortex neurons in vivo.' *Science.* 360(6395):1349–54. doi:10.1126/science.aao0862. PMID: 29930137; PMCID: PMC6366621

Deary, Ian J., Pattie, Alison, Starr, John M. (2013) 'The Stability of Intelligence From Age 11 to Age 90 Years: The Lothian Birth Cohort of 1921.' *Psychological Science*, vol. 24 issue 12, 2361–8

Lieberwirth, C., Wang, Z. (2012) 'The social environment and neurogenesis in the adult Mammalian brain.' *Front Hum Neurosci.* 6:118. doi:10.3389/fnhum.2012.00118

Kuhn, H. G., Toda, T., Gage, F. H. (2018) 'Adult Hippocampal Neurogenesis: A Coming-of-Age Story.' *J Neurosci.* 38(49):10401–10. doi:10.1523/JNEUROSCI.2144-18.2018. Epub 31 Oct 2018. PMID: 30381404; PMCID: PMC6284110

Chung, E. O. *et al.* (2020) 'The contribution of grandmother involvement to child growth and development: an observational study in rural Pakistan.' *BMJ Glob Health* 5, e002181. doi:10.1136/bmjgh-2019-002181

Rilling, J. K., Gonzalez, A., Lee, M. (2021) 'The neural correlates of grandmaternal caregiving.' *Proc Biol Sci.* 288(1963):20211997. doi:10.1098/rspb.2021.1997. Epub 17 Nov 2021. PMID: 34784762; PMCID: PMC8596004

Sear, R., Coall, D. (2011) 'How much does family matter? Cooperative breeding and the demographic transition.' *Popul Dev Rev.* 37, 81–112. doi:10.1111/j.1728-4457.2011.00379.x

Lehti, H., Erola, J., Tanskanen, A. O. (2019) 'Tying the extended family knot: grandparents' influence on educational achievement.' *Eur Sociol Rev.* 35, 29–48. doi:10.1093/esr/jcy044

Park, E. H. (2018) 'For Grandparents' Sake: the Relationship between Grandparenting Involvement and Psychological Well-Being.' *Ageing Int*;43(3):297–320. doi:10.1007/s12126-017-9320-8. Epub 16 Jan 2018. PMID: 30174357; PMCID: PMC6105248

Grandparents contribute to children's wellbeing: https://www.ox.ac.uk/research/research-impact/grandparents-contribute-childrens-wellbeing

Gonzalez, J., Anuncibay, R. (2008) 'Intergenerational grandparent/grandchild relations: the socioeducational role of grandparents.' *Educational Gerontology.* 34(1):67–88

Danielsbacka, M., Křenková, L. and Tanskanen, A. O. (2022) 'Grandparenting, health, and well-being: a systematic literature review.' *Eur J Ageing* https://doi.org/10.1007/s10433-021-00674-y

Men and Women – Is There a Cognitive Difference?

The Family Brain Games. BBC. https://www.bbc.co.uk/programmes/m00062jk

Woolley, A. W. *et al.* (2010) 'Evidence for a collective intelligence factor in the performance of human groups.' *Science* 330, 686–8

Riedl, C. *et al.* (2022) 'Quantifying collective intelligence in human groups [published correction appears in *Proc Natl Acad Sci USA.* 119(19):e2204380119]. *Proc Natl Acad Sci USA.* 2021;118(21):e2005737118. doi:10.1073/pnas.2005737118

Woolley, A. W., Aggarwal, I. (2020) 'Collective intelligence and group learning' in *Handbook of Group and Organizational Learning*, Argote, L., Levine, J. M., eds. Oxford University Press, London, UK, 491–506

IMAGEN Consortium (2021) 'The Human Brain Is Best Described as Being on a Female/Male Continuum: Evidence from a Neuroimaging Connectivity Study.' *Cerebral Cortex*, vol. 31, issue 6, 3021–33, https://doi.org/10.1093/cercor/bhaa408

'"Male" vs "female" brains: having a mix of both is common and offers big advantages – new research, 20 January 2021 https://theconversation.com/male-vs-female-brains-having-a-mix-of-both-is-common-and-offers-big-advantages-new-research-153242

Qiang, Luo, and Sahakian, Barbara J. (2022) 'Brain sex differences: the androgynous brain is advantageous for mental health and well-being.' *Neuropsychopharmacology: official publication of the American College of* Neuropsychopharmacology, vol. 47, 1 407–8. doi:10.1038/s41386-021-01141-z

Laying the Foundations for Collective Intelligence

Redhead, D., Power, E. A. (2022) 'Social hierarchies and social networks in humans.' *Philos Trans R Soc Lond B Biol Sci.* 377(1845):20200440. doi:10.1098/rstb.2020.0440. Epub 10 Jan 2022. PMID: 35000451; PMCID: PMC8743884

Haan, Ki-Won, Riedl, Christoph and Williams Woolley, Anita (2021) 'Discovering Where We Excel: How Inclusive Turn-Taking in Conversation Improves Team Performance.' In *Companion Publication of the 2021 International Conference on Multimodal Interaction (ICMI '21 Companion)*, 18–22 October 2021, Montréal, QC, Canada. ACM, New York, NY, USA, 8 pages. https://doi.org/10.1145/3461615.3485417

Larson, James Jr. (2013) *In Search of Synergy in Small Group Performance*. New York, NY: Psychology Press

Macmillan, Jean, Entin, Elliot E. and Serfaty, Daniel (2004) 'A framework for understanding the relationship between team structure and the communication necessary for effective team cognition.' In Salas, E., Fiore, S. M. and Cannon-Bowers, J., (eds),

Team Cognition: Process and Performance at the Inter- and Intra-Individual Level. Washington, DC, USA: APA

Engel, David *et al.* (2014) 'Reading the mind in the eyes or reading between the lines? Theory of mind predicts collective intelligence equally well online and face-to-face.' *PLoS One*, 9(12), article e115212.

Woolley, A. W. *et al.* (2010) 'Evidence for a collective intelligence factor in the performance of human groups.' *Science*, 330(6004), 686–8

Riedl, Christoph *et al.* (2021) 'Quantifying collective intelligence in human groups.' *Proceedings of the National Academy of Sciences*, 118(21)

Dunbar, R. I. *et al.* (2012) 'Social laughter is correlated with an elevated pain threshold.' *Proc Biol Sci.* 279(1731):1161–7. doi:10.1098/rspb.2011.1373. Epub 14 Sept 2011. PMID: 21920973; PMCID: PMC3267132

Kurtz, L. E., Algoe, S. B. (2015) 'Putting Laughter in Context: Shared Laughter as Behavioral Indicator of Relationship Well-Being.' *Pers Relatsh.*;22(4):573–90. doi:10.1111/pere.12095

Scott, Sophie, 'Why We Laugh.' TEDx talk, March 2015 https://www.ted.com/talks/sophie_scott_why_we_laugh?language=en

'The science of laughter', https://www.physoc.org/magazine-articles/the-science-of-laughter/ https://doi.org/10.36866/pn.103.34, Summer 2016, issue 103

'"Hyperscans" Show How Brains Sync as People Interact, Social neuroscientists ask what happens at the level of neurons when you tell someone a story or a group watches movies.' Lydia Denworth 10 April 2019, https://www.scientificamerican.com/article/hyperscans-show-how-brains-sync-as-people-interact

Read Montague, P. (2002) 'Hyperscanning: simultaneous fMRI during linked social interactions.' *NeuroImage.* 16 (4): 1159–64. doi:10.1006/nimg..1150. ISSN 1053-8119. PMID:12202103. S2CID: 15988039

Hasson, Uri *et al.* (2012). 'Brain-to-brain coupling: a mechanism for creating and sharing a social world.' *Trends in Cognitive Sciences.* 16 (2): 114–21. doi:10.1016/j.tics.2011.12.007. ISSN 1364-6613. PMC 3269540. PMID:22221820

Hu, Yi *et al.* (2018) 'Inter-brain synchrony and cooperation context in interactive decision making.' *Biological Psychology.* 133: 54–62. doi:

10.1016/j.biopsycho.2017.12.005. ISSN 1873-6246. PMID:29292232. S2CID: 46859640

Liu, Difei *et al.* (2018) 'Interactive Brain Activity: Review and Progress on EEG-Based Hyperscanning in Social Interactions.' *Frontiers in Psychology.* 9: 1862. doi:10.3389/fpsyg.2018.01862. ISSN 1664-1078. PMC 6186988. PMID:30349495

Babiloni, Fabio and Astolfi, Laura (2014) 'Social neuroscience and hyperscanning techniques: Past, present and future.' *Neuroscience & Biobehavioral Reviews. Applied Neuroscience: Models, methods, theories, reviews.* A Society of Applied Neuroscience (SAN) special issue. 44: 76–93. doi:10.1016/j.neubiorev.2012.07.006. ISSN 0149-7634. PMC 3522775. PMID:22917915

Leong, Victoria *et al.* (2017) 'Speaker gaze increases information coupling between infant and adult brains.' *PNAS* 114 (50) 13290–5, https://doi.org/10.1073/pnas.1702493114

Davidesco, Ido *et al.* (2019) 'Brain-to-brain synchrony between students and teachers predicts learning outcomes.' bioRxiv 644047; doi:https://doi.org/10.1101/644047

Davidesco, Ido *et al.* (2019) 'Brain-to-brain synchrony predicts long-term memory retention more accurately than individual brain measures.' doi:https://doi.org/10.1101/644047

Santamaria, L. *et al.* (2019) 'Emotional valence modulates the topology of the parent-infant inter-brain network.' *Neuroimage.* Doi:10.1016/j.neuroimage.2019.116341

Pan, Y. *et al.* (2021) 'Dual brain stimulation enhances interpersonal learning through spontaneous movement synchrony.' *Soc Cogn Affect Neurosci.*;16(1–2):210–221. doi:10.1093/scan/nsaa080

Valencia, Ana Lucía and Froese, Tom (2020) 'What binds us? Inter-brain neural synchronization and its implications for theories of human consciousness.' *Neuroscience of Consciousness*, vol. 2020, issue 1, niaa010, https://doi.org/10.1093/nc/niaa010

Hirsch, Joy *et al.* (2021) 'Interpersonal Agreement and Disagreement During Face-to-Face Dialogue: An fNIRS Investigation.' *Frontiers in Human Neuroscience*,14, https://www.frontiersin.org/article/10.3389/fnhum.2020.606397, doi:10.3389/fnhum.2020.606397

Hou, Y. *et al.* (2020) 'The averaged inter-brain coherence between the audience and a violinist predicts the popularity of violin performance.' *Neuroimage.* 211:116655. doi:10.1016/j.neuroimage. 2020.116655. Epub 18 Feb 2020. PMID: 32084565

Dotov, Dobromir *et al.* (2006) 'Collective music listening: Movement energy is enhanced by groove and visual social cues.' *Quarterly Journal of Experimental Psychology* vol. 74,6: 1037–53. doi:10.1177/ 1747021821991793

Osaka, Naoyuki *et al.* (2015) 'How Two Brains Make One Synchronized Mind in the Inferior Frontal Cortex: fNIRS-Based Hyperscanning During Cooperative Singing.' *Frontiers in Psychology* vol. 6, 1811. doi:10.3389/fpsyg.2015.01811

Lindenberger, U., Li, SC., Gruber, W. et al. Brains swinging in concert: cortical phase synchronization while playing guitar. BMC Neurosci 10, 22 (2009). https://doi.org/10.1186/1471-2202-10-22

Gao, J. *et al.* (2019) 'The neurophysiological correlates of religious chanting.' *Sci Rep* 9, 4262 https://doi.org/10.1038/s41598-019-40200-w

Jiang, J. *et al.* (2012) 'Neural synchronization during face-to-face communication.' *J Neurosci.* 32, 16064–9. 10.1523/JNEUROSCI.29 26-12.2012

Cui, X., Bryant, D. M., Reiss, A. L. (2012). 'NIRS-based hyperscanning reveals increased interpersonal coherence in superior frontal cortex during cooperation.' *Neuroimage* 59, 2430–7. 10.1016/j.neuroimage. 2011.09.003

Schofield, Timothy P.; Creswell, J. David; Denson, Thomas F. (2015). 'Brief mindfulness induction reduces inattentional blindness.' *Consciousness and Cognition.* 37: 63–70. doi:10.1016/j.concog. 2015.08.007. PMID:26320867

RocíoMartínez, Vivot *et al.* (2020) 'Meditation Increases the Entropy of Brain Oscillatory Activity.' *Neuroscience* vol. 431, 40–51 https:// doi.org/10.1016/j.neuroscience.2020.01.033

Lee, Darrin J. *et al.* 2018 'Review of the Neural Oscillations Underlying Meditation Front.' *Neurosci.* https://doi.org/10.3389/ fnins.2018.00178

Wegner, D. M., Erber, R., Raymond, P. (1991) 'Transactive memory in close relationships.' *J Pers Soc Psychol.* 61(6):923–9. doi:10.1037//0022-3514.61.6.923. PMID: 1774630

Wegner, D. M. (1987) 'Transactive Memory: A Contemporary Analysis of the Group Mind.' In: Mullen, B., Goethals, G. R. (eds) *Theories of Group Behavior* (Springer Series in Social Psychology), New York, USA: Springer

Hollingshead, A. B. (1998) 'Retrieval processes in transactive memory systems.' *Journal of Personality and Social Psychology,* 74(3), 659–71. https://doi.org/10.1037/0022-3514.74.3.659

Hollingshead, A. (2001). 'Cognitive interdependence and convergent expectations in transactive memory.' *Journal of Personality and Social Psychology.* 81 (6): 1080–9. doi:10.1037/0022-3514.81.6.1080. PMID:11761309

Hewitt, L. Y., Roberts, L. D. (2015) 'Transactive memory systems scale for couples: development and validation.' *Front Psychol.* 6:516. doi:10.3389/fpsyg.2015.00516. PMID: 25999873; PMCID: PMC4419599

Harris, Celia B. *et al.* (2021) 'It's not who you lose, it's who you are: Identity and symptom trajectory in prolonged grief.' *Current Psychology (New Brunswick, N.J.),* 1–11. doi:10.1007/s12144-021-02343-w

Sparrow, B., Liu, J, Wegner, D. M. (2011) 'Google effects on memory: cognitive consequences of having information at our fingertips.' *Science.* 333(6043):776–8. doi:10.1126/science.1207745. Epub 14 Jul 2011. PMID: 21764755

Zhang, Z.X. *et al.* (2007) 'Transactive memory system links work team characteristics and performance.' *J Appl Psychol.* 92(6):1722–30. doi:10.1037/0021-9010.92.6.1722. PMID: 18020808

Fiorenzato, Eleonora *et al.* 'Impact of COVID-19-lockdown and vulnerability factors on cognitive functioning and mental health in Italian population.' doi:https://doi.org/10.1101/2020.10.02.20205237 Pre-print; not yet peer-reviewed

RESONANCE Consortium. 'Impact of the COVID-19 Pandemic on Early Child Cognitive Development: Initial Findings in a Longitudinal Observational Study of Child Health.' medRxiv. Pre-print. doi:https://doi.org/10.1101/2021.08.10.21261846; this version posted 11 August 2021

Ingram, Joanne, Hand, Christopher J., Maciejewski, Greg. (2021) 'Social isolation during COVID-19 lockdown impairs cognitive function.' *Journal of Applied Cognitive Psychology.* https://doi.org/10.1002/acp.3821   https://onlinelibrary.wiley.com/doi/10.1002/acp.3821

Orben, A., Tomova, L., Blakemore, S. J. (2020) 'The effects of social deprivation on adolescent development and mental health.' *Lancet Child Adolesc Health.* 4(8):634–640. doi:10.1016/S2352-4642 (20)30186-3. Epub 12 June 2020. PMID: 32540024; PMCID: PMC7292584.

Zhang, S.X. *et al.* (2020) 'Unprecedented disruption of lives and work: Health, distress and life satisfaction of working adults in China one month into the COVID-19 outbreak.' *Psychiatry Research*, 288, 112958. https://doi.org/10.1016/j.psychres.2020.112958

Zunzunegui, M.-V. *et al.* (2003) 'Social networks, social integration, and social engagement determine cognitive decline in community-dwelling Spanish older adults.' *Journal of Gerontology*, 58B, S93–S100. https://doi.org/10.1093/geronb/58.2.s93

Cacioppo, J.T. *et al.* (2000) 'Lonely traits and concomitant physiological processes: The MacArthur social neuroscience studies.' *International Journal of Physiology*, 35(2–3), 143–54. https://doi.org/10.1016/s0167-8760(99)00049-5

Evans, I. E. M. *et al.* (2018) 'Social isolation, cognitive reserve, and cognition in healthy older people.' *PLoS One*, 13(8), e0201008. http://dx.doi.org/10.1371/journal.pone.0201008

Cacioppo, J. T., Hawkley, L. C., Thisted, R. A. (2010) 'Perceived social isolation makes me sad: 5-year cross-lagged analyses of loneliness and depressive symptomatology in the Chicago Health, Aging, and Social Relations Study.' *Psychol Aging.* 25(2):453–63. doi:10.1037/a0017216. PMID: 20545429; PMCID: PMC2922929.

Meltzer, H. *et al.* (2013) 'Feelings of loneliness among adults with mental disorder.' *Soc Psychiatry Psychiatr Epidemiol.* 48(1):5-13. doi:10.1007/s00127-012-0515-8. Epub 9 May 2012. Erratum in: *Soc Psychiatry Psychiatr Epidemiol.* 2015 Mar;50(3):503-4. PMID: 22570258

Silk, J. B. 'Evolutionary Perspectives on the Links Between Close

Social Bonds, Health, and Fitness.' In: Committee on Population; Division of Behavioral and Social Sciences and Education; National Research Council; Weinstein M., Lane M.A. (eds). *Sociality, Hierarchy, Health: Comparative Biodemography: A Collection of Papers.* Washington, D. C.: National Academies Press (US); 22 September 2014. 6. Available from: https://www.ncbi.nlm.nih.gov/books/NBK242452

Johnson, Zachary and Young, Larry (2015) 'Neurobiological mechanisms of social attachment and pair bonding.' *Current Opinion in Behavioral Sciences.* 3. 38–44. 10.1016/j.cobeha.2015.01.009

Cacioppo, J. T. *et al.* (2009) 'In the eye of the beholder: individual differences in perceived social isolation predict regional brain activation to social stimuli.' *J Cogn Neurosci.* 21(1):83–92. doi:10.1162/jocn.2009.21007. PMID: 18476760; PMCID: PMC2810252

Cacioppo, J.T., Chen, H.Y., Cacioppo, S. (2017) 'Reciprocal Influences Between Loneliness and Self-Centeredness: A Cross-Lagged Panel Analysis in a Population-Based Sample of African American, Hispanic, and Caucasian Adults.' *Pers Soc Psychol Bull.* 43(8):1125–35. doi:10.1177/0146167217705120. Epub 13 Jun 2017. PMID: 28903715

4. Intelligence at Work: Recruiting the Right Team

The Business of Escape Rooms and their worldwide success: https://www.marketwatch.com/story/the-weird-new-world-of-escape-room-businesses-2015-07-20

The Escape Room Challenge: How many PhDs does it take to break free? https://www.crypticevents.co.uk/blog.php?item=18

Setting Up Our Group for Success

Powell, J. L. *et al.* (2010) 'Orbital prefrontal cortex volume correlates with social cognitive competence.' *Neuropsychologia.* 48(12):3554–62. doi:10.1016/j.neuropsychologia.2010.08.004. Epub 14 Aug 2010. PMID: 20713074

Lewis, P. A. *et al.* (2011) 'Ventromedial prefrontal volume predicts understanding of others and social network size.' *Neuroimage.*

57(4):1624–9. doi:10.1016/j.neuroimage.2011.05.030. Epub 15 May 2011. PMID: 21616156

Silicon Valley and Apple Success was due to small community size fostering innovation: https://www.cnet.com/news/steve-wozniak-on-homebrew-computer-club/

The optimal group size and friendship groups: Professor Robin Dunbar and the Social Brain Hypothesis: as discussed in Critchlow, Hannah (2019) *The Science of Fate: The New Science of Who We Are – And How to Shape our Best Future*, London, UK: Hodder & Stoughton

Wright, N. D. *et al.* (2012) 'Testosterone disrupts human collaboration by increasing egocentric choices.' *Proc Biol Sci, 279 (1736)*, 2275-80. doi:10.1098/rspb.2011.2523

Mahmoodi, A., *et al.* (2015) 'Equality bias impairs collective decision-making across cultures.' *PROCEEDINGS OF THE NATIONAL ACADEMY OF SCIENCES OF THE UNITED STATES OF AMERICA, 112 (12)*, 3835-40. doi:10.1073/pnas.1421692112

Diversity Is the Key to Intelligent Groups

Peri, Giovanni, Shih, Kevin and Sparber, Chad (2015) 'STEM Workers, H-1B Visas, and Productivity in US Cities.' *Journal of Labor Economics*, vol. 33, no. S1. 'US High-Skilled Immigration in the Global Economy,' S225–55. The University of Chicago Press on behalf of the Society of Labor Economists and the NORC at the University of Chicago: http://www.jstor.org/stable/10.1086/679061

Page, Scott E. (2017) *The Diversity Bonus: How Great Teams Pay Off in the Knowledge Economy*. Princeton, NJ: Princeton University Press.

Page, Scott E. (2010) *Diversity and Complexity*. Princeton, NJ: Princeton University Press

Glennon, Britta (2020) 'How Do Restrictions on High-Skilled Immigration Affect Offshoring? Evidence from the H-1B Program.' Available at SSRN: https://ssrn.com/abstract=3547655 or http://dx.doi.org/10.2139/ssrn.3547655

https://www.techpolicy.com/Blog/March-2012/The-Case-for-Immigration.aspx

Ariely, D. *et al.* (2000) 'The effects of averaging subjective probability estimates between and within judges.' *J Exp Psychol Appl.* 6, 130–47. doi:10.1037/1076-898X.6.2.130

Johnson, T. R., Budescu, D. V., Wallsten, T. S. (2001) 'Averaging probability judgments: Monte Carlo analyses of asymptotic diagnostic value.' *J. Behav. Decis. Making* 14, 123–40. doi:10.1002/bdm.369

Migdał, P. *et al.* (2012) 'Information-sharing and aggregation models for interacting minds.' *J Math Psychol* 56, 417–26. doi:10.1016/j.jmp.2013.01.002

Krause, J., Ruxton, G. D., Krause, S. (2010) 'Swarm intelligence in animals and humans.' *Trends Ecol Evol* 25, 28–34. doi:10.1016/j.tree.2009.06.016

Couzin, I. D. (2009) 'Collective cognition in animal groups.' *Trends Cogn Sci.* 13, 36–43. doi:10.1016/j.tics.2008.10.002

Wuchty, Stefan, Jones, Benjamin F., Uzzi, Brian (2007) 'The increasing dominance of teams in production of knowledge.' 316(5827):1036–9. doi:10.1126/science.1136099. Epub 12 Apr 2007. PMID: 17431139. doi:10.1126/science.1136099

Uzzi, B. *et al.* (2013) 'Atypical combinations and scientific impact.' *Science*, 342(6157):468–72. doi:10.1126/science.1240474

Ahmadpoor, Mohammad and Jones, Benjamin F. (2019) 'Decoding team and individual impact in science and invention.' *PNAS* 116 (28) 13885–90

Wuchty, Stefan, Jones, Benjamin F., Uzzi, Brian. (2007) 'The increasing dominance of teams in production of knowledge.' 316(5827):1036–9. doi:10.1126/science.1136099. Epub 12 Apr 2007. PMID: 17431139. doi:10.1126/science.1136099

Uzzi, B. *et al.* (2013) 'Atypical combinations and scientific impact.' *Science*, 342(6157):468–72. doi:10.1126/science.1240474

Ahmadpoor, Mohammad and Jones, Benjamin F. (2019) 'Decoding team and individual impact in science and invention.' *PNAS* 116 (28) 13885–90

Jane Goodall's Wild Chimpanzees. PBS. (1996) Retrieved 28 July 2010: https://www.pbs.org/wnet/nature/jane-goodalls-wild-chimpanzees-introduction/1908

Goodall, Jane (1999) *Reason for Hope: A Spiritual Journey.* New York, NY: Warner Books

'From Top to Bottom, chimpanzee social hierarchy is amazing.' Brittany Cohen-Brown, 10 July 2018 https://news.janegoodall.org/2018/07/10/top-bottom-chimpanzee-social-hierarchy-amazing

Pusey, A. E., Schroepfer-Walker, K. (2013) 'Female competition in chimpanzees.' *Philos Trans R Soc Lond B Biol Sci.* 368(1631):20130077. doi:10.1098/rstb.2013.0077

Hemelrijk, C. K., Wantia, J., Isler, K. (2008) 'Female dominance over males in primates: self-organisation and sexual dimorphism.' *PLoS One.* 3(7):e2678.doi:10.1371/journal.pone.0002678.PMID:18628830; PMCID: PMC2441829

Izar, P. *et al.* (2021) 'Female emancipation in a male dominant, sexually dimorphic primate under natural conditions.' *PLoS One.* 16 (4):e0249039. doi:10.1371/journal.pone.0249039. PMID: 33872318; PMCID: PMC8055024

How Diverse Teams Work Best: Balancing the Skills

Robson, David (2019) *The Intelligence Trap: Revolutionise Your Thinking and Make Wiser Decisions* (first edition). London, UK: Hodder & Stoughton

PUZZLE SOLUTION

Jack (married)   Anne (not married)   George (not married)

OR

Jack (married)   Anne (married)   George (not married)

In either case, one married person will be looking at one unmarried person. The key to working this out is to accept ambiguity and consider the different solutions in order to arrive at the answer.

Gopnik, A. *et al.* (2017) 'Changes in cognitive flexibility and hypothesis search across human life history from childhood to adolescence to adulthood.' *Proc Natl Acad Sci USA.* 114(30):7892–9. doi:10.1073/pnas.1700811114

'Are adolescents more creative than adults?' Claire Stevenson, 13 January 2020 https://bold.expert/are-adolescents-more-creative-than-adults

Kleibeuker, S. W., de Dreu, C. K., Crone, E. A. (2013) 'The development of creative cognition across adolescence: distinct trajectories for insight and divergent thinking.' *Dev Sci*.16(1):2–12. doi:10.1111/j.1467-7687.2012.01176.x. Epub 8 Oct 2012. PMID: 23278922

Stevenson, C. E. *et al.* (2014) 'Training creative cognition: adolescence as a flexible period for improving creativity.' *Front Hum Neurosci*. 8:827. doi:10.3389/fnhum.2014.00827. PMID: 25400565; PMCID: PMC4212808

Crone, E. A., Dahl, R. E. (2012) 'Understanding adolescence as a period of social–affective engagement and goal flexibility.' *Nat Rev Neurosci*. 13, 636–50

Wu, Chi Hang *et al.* (2010) 'Age Differences in Creativity:Task Structure and Knowledge Base.' 321–6. Published online: *Creativity Research Journal* vol. 17, 2005, issue 4

Kleibeuker, S. W., de Dreu, C. K., Crone, E. A. (2013) 'The development of creative cognition across adolescence: distinct trajectories for insight and divergent thinking.' *Dev Sci*.;16 (1):2–12. doi:10.1111/j.1467-7687.2012.01176.x

Powell, J. L. *et al.* (2010) 'Orbital prefrontal cortex volume correlates with social cognitive competence.' *Neuropsychologia*. 48(12):3554–62. doi:10.1016/j.neuropsychologia.2010.08.004. Epub 14 Aug 2010. PMID: 20713074

Lewis, P. A. *et al.* (2011) 'Ventromedial prefrontal volume predicts understanding of others and social network size.' *Neuroimage*. 57(4):1624–9. doi:10.1016/j.neuroimage.2011.05.030. Epub 15 May 2011. PMID: 21616156

The optimal group size and friendship groups: Professor Robin Dunbar and the Social Brain Hypothesis: as discussed in Critchlow, Hannah (2019), *The Science of Fate: The New Science of Who We Are – And How to Shape our Best Future.* London, UK: Hodder & Stoughton

Draw on All the Talents: Neurodiversity

Nicolaou, N. *et al.* (2011) 'A polymorphism associated with entrepreneurship: evidence from dopamine receptor candidate genes.' *Small Bus Econ* 36, 151–5. https://doi.org/10.1007/s11187-010-9308-1

Wiklund, Johan, Patzelt, Holger, Dimov, Dimo (2016) 'Entrepreneurship and psychological disorders: How ADHD can be productively harnessed.' *Journal of Business Venturing Insights* 6:14 doi:10.1016/j.jbvi.2016.07.001

Sônego, M. *et al.* (2020) 'Exploring the association between attention-deficit/hyperactivity disorder and entrepreneurship.' *Braz J Psychiatry* S1516-44462020005018204. doi:10.1590/1516-4446-2020-0898

White, H. A. and Shah, P. (2006) 'Uninhibited imaginations: creativity in adults with attention-deficit/hyperactivity disorder.' *Pers Individ Differ.* 40,1121–31. doi:10.1016/j.paid.2005.11.007

White, H. A. and Shah, P. (2011) 'Creative style and achievement in adults with attention-deficit/hyperactivity disorder.' *Pers Individ Differ.* 50, 673–7. doi:10.1016/j.paid.2010.12.015

Guilford, J. P. (1967) *The Nature of Human Intelligence*. New York, NY: McGraw-Hill

Shpigler, Hagai Y. *et al.* (2017) 'Evolutionary conservation of autism genes.' *Proceedings of the National Academy of Sciences* 114 (36) 9653–8; doi:10.1073/pnas.1708127114

'Is my autism a superpower?' 3 Nov 2019 https://www.theguardian.com/society/2019/nov/03/is-autism-a-superpower-greta-thunberg-and-others-think-it-can-be

'These major tech companies are making autism hiring a priority.' https://www.monster.com/career-advice/article/autism-hiring-initiatives-tech

'Greta Thunberg: Why She Called Aspergers Her Superpower.' https://www.forbes.com/sites/brucelee/2019/09/27/greta-thunberg-why-she-called-aspergers-her-superpower/?sh=174fc4ce4101

Critchlow, H. M. (2007). 'The Role of Dendritic Spine Plasticity in Schizophrenia' (doctoral thesis, University of Cambridge)

Kaufman, S. B., Paul, E. S. (2014) 'Creativity and schizophrenia spectrum disorders across the arts and sciences.' *Front Psychol.* 5:1145. doi:10.3389/fpsyg.2014.01145

Hilker, R. *et al.* (2018) 'Heritability of Schizophrenia and Schizophrenia Spectrum Based on the Nationwide Danish Twin Register.' *Biol Psychiatry.* 83(6):492–8. doi:10.1016/j.biopsych.2017.08.017. Epub 1 Sep 2017. PMID: 28987712

Kyaga, S. *et al.* (2013) 'Mental illness, suicide and creativity: 40-year prospective total population study.' *J Psychiatr Res.* 47(1):83–90. doi:10.1016/j.jpsychires.2012.09.010. Epub 9 Oct 2012. PMID: 23063328

Kyaga, S. *et al.* (2011) 'Creativity and mental disorder: family study of 300,000 people with severe mental disorder.' *Br J Psychiatry.* 199(5):373–9. doi:10.1192/bjp.bp.110.085316. Epub 8 Jun 2011. PMID: 21653945

Johns, L. C. and van Os, J. (2001) 'The continuity of psychotic experiences in the general population.' *Clin Psychol Rev.* 21(8):1125–41

Olfson, M. *et al.* (2002) 'Psychotic symptoms in an urban general medicine practice.' *Am J Psychiatry.* 159(8):1412–9

Kendler, K. S. *et al.* (1996) 'Lifetime prevalence, demographic risk factors, and diagnostic validity of nonaffective psychosis as assessed in a US community sample.' The National Comorbidity Survey. *Arch Gen Psychiatry.* 53(11):1022–31

Tien, A. Y. (1991) 'Distributions of hallucinations in the population.' *Soc Psychiatry Psychiatr Epidemiol.* 26(6):287–92

Temmingh, H. *et al.* (2011) 'The prevalence and correlates of hallucinations in a general population sample: findings from the South African Stress and Health Study.' *Afr J Psychiatry (Johannesbg)* 14(3):211–17. doi:10.4314/ajpsy.v14i3.4

Coid, J. W. *et al.* (2008) 'Raised incidence rates of all psychoses among migrant groups: findings from the East London first episode psychosis study.' *Arch Gen Psychiatry.* 65(11):1250–8. doi:10.1001/archpsyc.65.11.1250. Erratum in: *Arch Gen Psychiatry.* 2009 Feb;66(2):161. PMID: 18981336

The Family Brain Games. BBC. https://www.bbc.co.uk/programmes/m00062jk

Bahrami, B. *et al.* (2010). 'Optimally interacting minds.' *Science*, 329 (5995), 1081–5. doi:10.1126/science.1185718

5. Harnessing Brain Power: Strategies for Smart Teams

Leading for Collective Intelligence

https://www.businessinsider.com.au/larry-page-the-untold-story-2014-4?page=2&r=US&IR=T

Tsvetkova, Milena *et al.* (2017) 'Even Good Bots Fight: The case of Wikipedia.' *PLoS One* https://doi.org/10.1371/journal.pone.0171774

Power and Consensus

Hare, R. D., Neumann, C. S. (2008) 'Psychopathy as a clinical and empirical construct.' *Annu Rev Clin Psychol.* 4:217–46.doi:10.1146/annurev.clinpsy.3.022806.091452. PMID: 18370617

Cima, M., Tonnaer, F., Hauser, M. D. (2010) 'Psychopaths know right from wrong but don't care.' *Soc Cogn Affect Neurosci.* 5(1):59–67. doi:10.1093/scan/nsp051

https://www.telegraph.co.uk/news/2016/09/13/1-in-5-ceos-are-psychopaths-australian-study-finds

https://www.forbes.com/sites/jackmccullough/2019/12/09/the-psychopathic-ceo/?sh=314ff015791e

Babiak, Paul and Hare, Robert D. (2019) *Snakes in Suits: Understanding and Surviving the Psychopaths in Your Office* (revised edition). Harper Business

Landay, K., Harms, P. D., Credé, M. (2019) 'Shall we serve the dark lords? A meta-analytic review of psychopathy and leadership.' *Journal of Applied Psychology*, *[s.l.]*, vol. 104, no. 1, *Leadership*, 183–96. doi:10.1037/apl0000357. Available from: https://search-ebscohost-com.ezp.lib.cam.ac.uk/login.aspx?direct=true&db=pdh&AN=2018-51219-001&site=ehost-live&scope=site. Accessed 22 July 2021

All about wolves: http://teacher.scholastic.com/wolves/gabout3.htm#:~:text=A%20wolf%20pack%20has%20a,to%20eat%20first%20at%20kills

Sumpter, David J.T. *et al.* (2008) 'Consensus Decision Making by Fish.' *Current Biology* 18, 1773–7

Jolles, Jolle W. 'Group-level patterns emerge from individual speed as revealed by an extremely social robotic fish.' doi:https://doi.org/10.1101/2020.06.10.143883. June 2020

Couzin I. D., *et al.* 'Effective leadership and decision-making in animal groups on the move. Nature. 2005 Feb 3;433(7025):513-6. doi:10.1038/nature03236

Dyer, John R. G. (2008) 'Leadership, consensus decision making and collective behaviour in humans.' *Philosophical Transactions of the Royal Society B: Biological Sciences* vol. 364, issue 1518. https://doi.org/10.1098/rstb.2008.0233

Krause, J. *et al.* (2021) 'Collective rule-breaking.' *Trends Cogn Sci.* 25(12):1082–95. doi:10.1016/j.tics.2021.08.003. Epub 4 Sep 2021. PMID: 34493441

Anicich, Eric M., Swaab, Roderick I. and Galinsky, Adam D. 'Hierarchical cultural values predict success and mortality in high-stakes teams.' *PNAS*, www.pnas.org/cgi/doi/10.1073/pnas.1408800112

The Myth of the Psycho CEO and the Pro-social Alternative

Anderson, Cameron *et al.* (2020) 'People with disagreeable personalities (selfish, combative, and manipulative) do not have an advantage in pursuing power at work.'

117 (37) 22780-22786 https://doi.org/10.1073/pnas.2005088117

Campbell-Meiklejohn *et al,* 'How the Opinion of Others Affects Our Valuation of Objects.' *Current Biology*, 17 June 2010. doi 10.1016/j.cub.2010.04.055

Woolley, A. W. *et al.* (2010) 'Evidence for a collective intelligence factor in the performance of human groups.' *Science* 330(6004)

Riedl, C. *et al.* (2021) 'Quantifying collective intelligence in human groups.' *Proc Natl Acad Sci USA.* 118(21):e2005737118. doi:10.1073/pnas.2005737118. PMID: 34001598; PMCID: PMC8166150

Wright, N. D. *et al.* (2012) 'Testosterone disrupts human collaboration by increasing egocentric choices.' *Proc Biol Sci.* 279(1736):2275–80. doi:10.1098/rspb.2011.2523. Epub 1 Feb 2012. PMID: 22298852; PMCID: PMC3321715

Decety, J. *et al.* (2016) 'Empathy as a driver of prosocial behaviour: highly conserved neurobehavioural mechanisms across species.' *Philos Trans R Soc Lond B Biol Sci.* 371(1686):20150077. doi:10.1098/rstb.2015.0077

Sharif, K. (2019) 'Transformational leadership behaviours of women in a socially dynamic environment.' *International Journal of Organizational Analysis*, vol. 27, no. 4, 1191–1217. https://doi.org/10.1108/IJOA-12-2018-1611

Suranga Silva, D. A. C., Mendis, B.A.K.M. (2017) 'Male vs Female Leaders: Analysis of Transformational, Transactional & Laissez-faire Women Leadership Styles.' *European Journal of Business and Management* www.iiste.org ISSN 2222-1905 (Paper) ISSN 2222-2839 (online) vol. 9, no. 9

'Research: Women Score Higher Than Men in Most Leadership Skills.' *Harvard Business Review*, 25 June 2019, https://hbr.org/2019/06/research-women-score-higher-than-men-in-most-leadership-skills

'Women CEOs: Why So Few?' *Harvard Business Review*, 21 December 2009, https://hbr.org/2009/12/women-ceo-why-so-few

Harvard Business Review 2019: https://hbr.org/2019/11/the-best-performing-ceos-in-the-world-2019

Yang, Y., Chawla, N. V., Uzzi, B. (2019) 'A network's gender composition and communication pattern predict women's leadership success.' *Proc Natl Acad Sci USA.* 116(6):2033–38. doi:10.1073/pnas.1721438116. Epub 22 Jan 2019. Erratum in: *Proc Natl Acad Sci USA.* 2019 April 116(14):7149. PMID: 30670641; PMCID: PMC6369753

'Controversy over the role of same gender mentorship.' RETRACTED ARTICLE: 'The association between early career informal mentorship in academic collaborations and junior author performance.' *Nat Commun* 11, 5855 (2020). https://doi.org/10.1038/s41467-020-19723-8

Bryant, John (2005) *3:59.4: The Quest to Break the 4 Minute Mile* (international edition). Arrow.

Haas, Tanner (2019) 'The STORY Of Roger Bannister: The Power Of Self-Belief.' LinkedIn, 12 July https://www.linkedin.com/pulse/story-roger-bannister-power-self-belief-tanner-haas

Group Dynamics, Dominance and Turn-taking

Syed, Matthew (2019) *Rebel Ideas: The Power of Diverse Thinking*. London, UK: John Murray

Anderson, Cameron and Kennedy, Jessica (2017) 'Hierarchical rank and principled dissent: How holding higher rank suppresses objection to unethical practices.' Organizational Behavior and Human Decision Processes 139:30–49 doi:10.1016/j.obhdp.2017.01.002

Taylor, William C. (2012) *Practically Radical: Not-So-Crazy Ways to Transform Your Company, Shake Up Your Industry, and Challenge Yourself.* William Morrow Paperbacks.

Steering Away From Groupthink

Page, Scott E. (2017) *The Diversity Bonus: How Great Teams Pay Off in the Knowledge Economy*. Princeton, NJ: Princeton University Press

Page, Scott E. (2010) *Diversity and Complexity*. Princeton, NJ: Princeton University Press

Syed, Matthew (2019) *Rebel Ideas: The Power of Diverse Thinking*. London, UK: John Murray

Surowiecki, James (2005) *The Wisdom of Crowds*. Anchor Books

Litcanua, Marcela (2015) 'Brain-Writing Vs. Brainstorming Case Study For Power Engineering Education.' *Procedia – Social and Behavioral Sciences*, 191

Resilient Brains Are Much More Intelligent

'The Mind and Mental Health: How Stress Affects the Brain' https://www.tuw.edu/health/how-stress-affects-the-brain/#:~:text=It%20can%20disrupt%20synapse%20regulation,responsible%20for%20memory%20and%20learning

Roiser J. P., Sahakian, B. J. (2013) 'Hot and cold cognition in depression.' *CNS Spectr.* 18 (3): 139–49. doi:10.1017/S1092852913000072. PMID:23481353. S2CID: 34123889

Nord, C. L. *et al.* (2020) 'The neural basis of hot and cold cognition in depressed patients, unaffected relatives, and low-risk healthy controls: An fMRI investigation.' *Journal of Affective Disorders* vol. 274, 389–98. doi:10.1016/j.jad.2020.05.022

Kuhn, H. G., Dickinson-Anson, H., Gage, F. H. (1996) 'Neurogenesis in the dentate gyrus of the adult rat: age-related decrease of neuronal progenitor proliferation.' *J Neurosci* 16:2027–33. 10.1523/ JNEUROSCI.16-06-02027

Kuhn, H. G., Toda, T., Gage, F. H. (2018) 'Adult Hippocampal Neurogenesis: A Coming-of-Age Story.' *J Neurosci.* 38(49):10401– 10410. doi:10.1523/JNEUROSCI.2144-18. Epub 31 Oct 2018. PMID: 30381404; PMCID: PMC6284110

Extinction Rebellion about us: https://extinctionrebellion.uk/ the-truth/about-us

Pushkarna, Akshit (2021) 'Why social interaction is essential to drive innovation.' HRKatha, 9 September www.hrkatha.com/features/ why-social-interaction-is-essential-to-drive-innovation

Liu, Xueyuan *et al.* (2017) 'The impact of informal social interaction on innovation capability in the context of buyer-supplier dyads.' *Journal of Business Research*, vol. 78, 314–22, ISSN 0148-2963, https:// doi.org/10.1016/j.jbusres.2016.12.027. https://www.sciencedirect. com/science/article/pii/S0148296317300723

Russo, Francine (2022) 'The Personality Trait "Intolerance of Uncertainty"

Causes Anguish during COVID; High levels of it have put people at risk of emotional problems.' *Scientific American*, 14 February https:// www.scientificamerican.com/article/the-personality-trait- intolerance-of-uncertainty-causes-anguish-during-covid

Mofrad, Layla *et al.* (2020) 'Making friends with uncertainty: experiences of developing a transdiagnostic group intervention targeting intolerance of uncertainty in IAPT. Feasibility, acceptability and implications.' Cognitive Behaviour Therapist , vol. 13, e49, doi:https:// doi.org/10.1017/S1754470X20000495

12-Point Tolerance of Uncertainty Test https://www.midss.org/ content/intolerance-uncertainty-scale-short-form-ius-12

Tolerance of Uncertainty Test: https://jennifershannon.com/ wp-content/uploads/2020/01/DontFeedMonkeyMind_ worksheets.pdf

Smart Ways with Conflict

Gallo, Amy (2018) 'Why We Should Be Disagreeing More at Work', HBR, 3 January https://hbr.org/2018/01/why-we-should-be-disagreeing-more-at-work

Kilduff, G. J., Willer, R. and Anderson, C. (2016). 'Hierarchy and its discontents: Status disagreement leads to withdrawal of contribution and lower group performance.' *Organization Science,* 27(2), 373–390. https://doi.org/10.1287/orsc.2016.1058

Wolpert, Daniel, 'The Real Reason for Brains (is to move).' TED talk, November 2011 https://www.ted.com/talks/daniel_wolpert_the_real_reason_for_brains/transcript?language=en

Neuroscientist Daniel Wolpert starts from a surprising premise: the brain evolved, not to think or feel, but to control movement. In this entertaining, data-rich talk he gives us a glimpse into how the brain creates the grace and agility of human motion:

Boost the Range of Emotional Intelligence

Colman, A. (2008) *A Dictionary of Psychology* (third edition). Oxford University Press.

Goleman, D. (1998) 'What Makes a Leader?' *Harvard Business Review.* 76: 92–105

Beldoch, Michael and Davitz, Joel Robert. (1976) *The Communication of Emotional Meaning.* Westport, Conn.: Greenwood Press, 39

Neumann, R. and Strack, F. (2000) '"Mood contagion": the automatic transfer of mood between persons.' *J Pers Soc Psychol.* 79(2):211–23. doi:10.1037//0022-3514.79.2.211. PMID: 10948975

Isabella, Giuliana and Carvalho, Hamilton C. (2016) 'Emotional Contagion and Socialization', in *Emotions, Technology, and Behaviors;* Elsevier.

Fowler, J. H., Christakis, N. A. (2008) 'Dynamic spread of happiness in a large social network: longitudinal analysis over 20 years in the Framingham Heart Study.' *BMJ*; 337: a2338 doi:https://doi.org/10.1136/bmj.a2338

Guillory, J., *et al.* (2011) 'Upset now? Emotion contagion in distributed groups.' *Proc ACM CHI Conf on Human Factors in Computing Systems* (Association for Computing Machinery, New York), 745–8

Kramer, Adam D. I., Guillory, Jamie E. and Hancock, Jeffrey T. (2014) 'Experimental evidence of massive-scale emotional contagion through social networks.' *PNAS* 111 (24) 8788–90; https://doi.org/10.1073/pnas.1320040111

Carnevale, P. J. and Isen, A. M. (1986) 'The influence of positive affect and visual access on the discovery of integrative solutions in bilateral negotiation.' *Organizational Behavior and Human Decision Processes,* 37(1), 1–13. https://doi.org/10.1016/0749-5978(86)90041-5

Barsade, Sigal G., Coutifaris, Constantinos G. V., Pillemer, Julianna (2018) 'Emotional contagion in organizational life.' Research in Organizational Behavior, vol. 38, 137–151

Iacoboni, Marco (2008) *Mirroring People: The New Science of How We Connect with Others* (first edition). Farrar, Straus and Giroux

Prochazkova, Eliska and Kret, Mariska E. (2017) 'Connecting minds and sharing emotions through mimicry: A neurocognitive model of emotional contagion.' *Neuroscience & Biobehavioral Reviews,* vol. 80, 99–114.

Christov-Moore, Leonardo, Conway, Paul, Iacoboni, Marco (2017) 'Deontological Dilemma Response Tendencies and Sensorimotor Representations of Harm to Others.' *Frontiers in Integrative Neuroscience* doi:10.3389/fnint.2017.00034

Dimberg, Ulf, Thunberg, Monika, Elmehed, Kurt (2000) 'Unconscious Facial Reactions to Emotional Facial Expressions.' Brief Report. Find in PubMed https://doi.org/10.1111/1467-9280.00221

Barsade, Sigal and O'Neill, Olivia A. (2016) 'Leadership & Managing People: Manage Your Emotional Culture'. *Harvard Business Review,* January–February

Young, Emma (2020) 'When a Smile is Not a Smile: what our facial expression really means.' *New Scientist* https://www.newscientist.com/article/mg24532690-900-when-a-smile-is-not-a-smile-what-our-facial-expressions-really-mean

Gino, Francesca, Ayal, Shahar, Ariely, Dan (2009) 'Contagion and Differentiation in Unethical Behavior: The Effect of One Bad Apple on the Barrel.' Research Article, *Psychological Science,* vol. 20, issue 3, 393–8 https://doi.org/10.1111/j.1467-9280.2009.02306.x

Barsade, Sigal G., Coutifaris, Constantinos G. V., Pillemer, Julianna (2018) 'Emotional contagion in organisational life.', vol. 38, 137–51. https://doi.org/10.1016/j.riob.2018.11.005

Friedman, Howard S. *et al.* (1980) 'Understanding and Assessing Nonverbal Expressiveness: The Affective Communication Test.' *Journal of Personality and Social Psychology*, vol. 39, no 2, 333–51

Neumann, R., Strack, F. (2000) '"Mood contagion": the automatic transfer of mood between persons.' *J Pers Soc Psychol.* 79(2):211–23. doi:10.1037//0022-3514.79.2.211. PMID: 10948975

Isabella, Giuliana and Carvalho, Hamilton C. (2016) 'Emotional Contagion and Socialization in Emotions, Technology, and Behaviors.' Elsevier

Guillory J. *et al.* (2011) 'Upset now? Emotion contagion in distributed groups.' *Proc ACM CHI Conf on Human Factors in Computing Systems* (Association for Computing Machinery, New York), 745–8

Kramer, Adam D. I., Guillory, Jamie E. and Hancock, Jeffrey T. (2014) 'Experimental evidence of massive-scale emotional contagion through social networks.' *PNAS* 111 (24) 8788–90 https://doi.org/10.1073/pnas.1320040111

Carnevale, P. J. and Isen, A. M. (1986) 'The influence of positive affect and visual access on the discovery of integrative solutions in bilateral negotiation.' *Organizational Behavior and Human Decision Processes,* 37(1), 1–13. https://doi.org/10.1016/0749-5978(86)90041-5

Barsade, Sigal G., Coutifaris, Constantinos G. V., Pillemer, Julianna (2018) 'Emotional contagion in organizational life.' Research in Organizational Behavior, vol. 38, 137–51.

Grant, A. M. and Gino, F. (2010) 'A little thanks goes a long way: Explaining why gratitude expressions motivate prosocial behavior.' *Journal of Personality and Social Psychology,* 98(6), 946–55. https://doi.org/10.1037/a0017935

Anicich, E. M. *et al.* 'Powerful and Ungrateful: Why Power Reduces Gratitude Expression.' (under review at OBHDP)

Dweck, Carol (2017) *Mindset: Changing The Way You Think To Fulfil Your Potential* (updated edition). Little, Brown Book Group.

Von Hippel, William, Ronay, Richard, Baker, Ernest (2016) 'Quick Thinkers Are Smooth Talkers: Mental Speed Facilitates Charisma.'

Psychol Sci 27(1):119–22. doi:10.1177/0956797615616255. Epub 30 Nov 2015

Schjoedt U. *et al.* (2011) 'The power of charisma – perceived charisma inhibits the frontal executive network of believers in intercessory prayer.' *Soc Cogn Affect Neurosci*.;6(1):119–27. doi:10.1093/scan/nsq023

Greiser, Christian *et al.*(2020) 'Tap Your Company's Collective Intelligence with Mindfulness.' bcg.com, 5 February https://www.bcg.com/publications/2020/tap-your-company-collective-intelligence-with-mindfulness

6. Intuitive Intelligence – The Next Great Untapped Skill

Jung, Carl G. (1971) *Psychological Types*. Princeton, New Jersey: Princeton University Press

https://www.forbes.com/sites/brucekasanoff/2017/02/21/intuition-is-the-highest-form-of-intelligence/#5e89f5f33860

https://kasanoff.com/blog/2020/5/18/intuition-is-the-highest-form-of-intelligence

Samples, Bob (1976) *The Metaphoric Mind: A Celebration of Creative Consciousness* Addison Wesley Longman Publishing Co

Pilard, Nathalie (2018) 'C. G. Jung and intuition: from the mindscape of the paranormal to the heart of psychology.' *Psychol* 63(1):65–84. doi:10.1111/1468-5922.12380. doi:10.1111/1468-5922.12380

Intuition Is a Superpower

Lufityanto, Galang, Donkin, Chris, Pearson, Joel (2016) 'Measuring Intuition: Nonconscious Emotional Information Boosts Decision Accuracy and Confidence.' *Psychol Sci*, 27(5):622–34. doi:10.1177/0956797616629403. Epub 6 April 2016

'"Continuous flash suppression." Vision Sciences Society Annual Meeting Abstract.' (2004) Suchiya, Naotsugu; Koch, Christof. *Journal of Vision*, vol. 4, 61. doi:https://doi.org/10.1167/4.8.61

Boucsein, Wolfram (2012) *Electrodermal Activity*. Springer Science + Business Media, 4

Vlassova, Alexandra, Donkin, Chris, Pearson, Joel (2014) 'Unconscious information changes decision accuracy but not confidence.' *Proc*

Natl Acad Sci USA 11 Nov;111(45):16214-8. doi:10.1073/pnas.1403619111. Epub 27 Oct 2014

Critchlow, Hannah (2018) 'Conscious Awareness and the Case of Patient DB' in *Consciousness: A Ladybird Expert Book* (The Ladybird Expert Series 29), London, UK: Michael Joseph

Mannes, A. E., Soll, J. B. and Larrick, R. P. (2014 'The wisdom of select crowds.' *Journal of Personality and Social Psychology*, 107(2), 276–99. https://doi.org/10.1037/a0036677

Palley, Asa B. and Soll, Jack B. (2019) 'Extracting the Wisdom of Crowds When Information Is Shared.' Published online 21 Feb 2019 https://doi.org/10.1287/mnsc.2018.3047

Kandasamy, Narayanan *et al.* (2016) 'Interoceptive Ability Predicts Survival on a London Trading Floor.' *Scientific Reports*, 6: 32986. doi:10.1038/srep32986

Updating the Wisdom of Crowds

Galton, F. (1907) 'Vox Populi.' *Nature* 75, 450–1. https://doi.org/10.1038/075450a0

Wallis, K. F. (2014) 'Revisiting Francis Galton's forecasting competition.' *Statistical Science,* 29, 420-4. doi:10.1214/14-STS468.

Mannes, A. E., Soll, J. B. and Larrick, R. P. (2014) 'The wisdom of select crowds.' *Journal of Personality and Social Psychology*, 107(2), 276–99. https://doi.org/10.1037/a0036677

Palley, Asa B. and Soll, Jack B. (2019) 'Extracting the Wisdom of Crowds When Information Is Shared.' Published online 21 Feb 2019 https://doi.org/10.1287/mnsc.2018.3047

Syed, Matthew (2019) *Rebel Ideas: The Power of Diverse Thinking.* London, UK: John Murray

Surowiecki, James (2005) *The Wisdom of Crowds.* Anchor Books

Gigerenzer, Gerd (2008) *Gut Feelings* (reprint edition). Penguin Books

Gladwell, Malcolm (2005) *Blink* (new edition) Little, Brown and Company

When Is a Hunch Just a Prejudice?

Starr, Douglas (2020) 'The Bias Detective.' *Science.com*, 26 March doi:10.1126/science.abb9022, https://www.science.org/content/

REFERENCES

article/meet-psychologist-exploring-unconscious-bias-and-its-tragic-consequences-society

Golby, A. J. *et al.* (2001) 'Differential responses in the fusiform region to same-race and other-race faces.' *Nat Neurosci.* 4(8):845–50. doi:10.1038/90565. PMID: 11477432

Hughes, B. L. *et al.* (2019) 'Neural adaptation to faces reveals racial outgroup homogeneity effects in early perception.' *Proc Natl Acad Sci USA.* 116(29):14532–7. doi:10.1073/pnas.1822084116. Epub 1 Jul 2019. PMID: 31262811; PMCID: PMC6642392

Golarai, G. *et al.* (2021) 'The development of race effects in face processing from childhood through adulthood: neural and behavioral evidence.' *Dev Sci.* 24(3):e13058. doi:10.1111/desc.13058. Epub 5 Dec 2020. PMID: 33151616

Voigt, R. *et al.* (2017) 'Language from police body camera footage shows racial disparities in officer respect.' *Proc Natl Acad Sci USA.* 114(25):6521–6. doi:10.1073/pnas.1702413114. Epub 5 Jun 2017. PMID: 28584085; PMCID: PMC5488942

Agarwal, Pragya (2020) 'What do unconscious bias tests really reveal about racism?' *New Scientist*, 26 August https://www.newscientist.com/article/mg24732973-400-what-do-unconscious-bias-tests-really-reveal-about-racism/#ixzz6WfV9LoXD

Ross, C. T. (2015) 'A Multi-Level Bayesian Analysis of Racial Bias in Police Shootings at the County-Level in the United States, 2011–2014.' *PLoS One* 10(11): e0141854. https://doi.org/10.1371/journal.pone.0141854

Eberhardt, J. L. *et al.* (2004) 'Seeing black: race, crime, and visual processing.' *J Pers Soc Psychol.* 87(6):876–93. doi:10.1037/0022-3514.87.6.876. PMID: 15598112

Garfinkel, S. N., Critchley, H. D. (2016) 'Threat and the Body: How the Heart Supports Fear Processing.' *Trends Cogn Sci.* 20(1):34–46. doi:10.1016/j.tics.2015.10.005

Critchley, H. D. *et al.* (2004) 'Neural systems supporting interoceptive awareness.' *Nat Neurosci.* 7(2):189–95. doi:10.1038/nn1176

Azevedo, R. *et al.* (2017). 'Cardiac afferent activity modulates the expression of racial stereotypes.' *Nat Commun* 8, 13854 https://doi.org/10.1038/ncomms13854

Remmers, Carina and Johannes Michalak (2016) 'Losing Your Gut Feelings. Intuition in Depression.' *Front Psychol.* 7:1291. doi:10.3389/fpsyg.2016.01291. eCollection

Khalsa, S. S. *et al.* 'Interoception and Mental Health: A Roadmap.' *Biol Psychiatry Cogn Neurosci Neuroimaging.* 2018;3(6):501–513. doi:10.1016/j.bpsc.2017.12.004

Garfinkel, Sarah N. *et al.* (2015) 'Knowing your own heart: distinguishing interoceptive accuracy from interoceptive awareness.' *Biol Psychol* 104:65–74. doi:10.1016/j.biopsycho.2014.11.004. Epub 20 Nov 2014

Critchley, Hugo D. and Garfinkel, Sarah N. (2017) 'Interoception and emotion.' *Current Opinion in Psychology*, 17. 7–14. ISSN 2352-250X

Ewing, Donna L. *et al.* (2017) 'Sleep and the heart: interoceptive differences linked to poor experiential sleep quality in anxiety and depression.' *Biological Psychology*, 127. 163–72. ISSN 0301-0511

Garfinkel, Sarah N. *et al.* (2016) 'Discrepancies between dimensions of interoception in autism: implications for emotion and anxiety.' *Biological Psychology*, 114. 117–26. ISSN 0301-0511

Remmers, C. *et al.* (2015) 'Impaired intuition in patients with major depressive disorder.' *Br J Clin Psychol.* 54(2):200–13. doi:10.1111/bjc.12069. Epub 11 Oct 2014

Can We Learn to Be Super-sensors?

Lufityanto, Galang, Donkin, Chris, Pearson, Joel (2016) 'Measuring Intuition: Nonconscious Emotional Information Boosts Decision Accuracy and Confidence.' *Psychol Sci*, 27(5):622–34. doi:10.1177/0956797616629403. Epub 6 April 2016

'"Continuous flash suppression." Vision Sciences Society Annual Meeting Abstract.' (2004) Suchiya, Naotsugu; Koch, Christof. *Journal of Vision*, vol. 4, 61. doi:https://doi.org/10.1167/4.8.61

Vlassova, Alexandra, Donkin, Chris, Pearson, Joel (2014) 'Unconscious information changes decision accuracy but not confidence.' *Proc Natl Acad Sci USA* 11 Nov;111(45):16214-8. doi:10.1073/pnas.1403619111. Epub 27 Oct 2014

'Aligning Dimensions of Interoceptive Experience (ADIE) to prevent development of anxiety disorders in autism.' https://www.mqmentalhealth.org/research/aligning-dimensions-of-interoceptive-experience-adie-to-prevent-development-of-anxiety-disorders-in-autism

Quadt, L. *et al.* (2021). 'Interoceptive training to target anxiety in autistic adults (ADIE): A single-center, superiority randomized controlled trial.' *EClinicalMedicine*, 39, 101042. https://doi.org/10.1016/j.eclinm.2021.101042

Does Intuitive Intelligence Work Remotely?

Fiorenzato, Eleonora *et al.* 'Impact of COVID-19-lockdown and vulnerability factors on cognitive functioning and mental health in Italian population:' doi:https://doi.org/10.1101/2020.10.02.20205237 Pre-print; not yet peer-reviewed

Ingram, Joanne, Hand, Christopher J., Maciejewski, Greg. (2021) 'Social isolation during COVID-19 lockdown impairs cognitive function.' *Journal of Applied Cognitive Psychology.* https://doi.org/10.1002/acp.3821 https://onlinelibrary.wiley.com/doi/10.1002/acp.3821

Orben, A., Tomova, L., Blakemore, S. J. (2020) 'The effects of social deprivation on adolescent development and mental health.' *Lancet Child Adolesc Health.* 4(8):634–640. doi:10.1016/S2352-4642(20)30186-3. Epub 12 June 2020. PMID: 32540024; PMCID: PMC7292584

Aczel, B. *et al.* (2021) 'Researchers working from home: Benefits and challenges.' *PLoS One* 16(3): e0249127. https://doi.org/10.1371/journal.pone.0249127

Tseng, P. H. *et al.* (2018) 'Interbrain cortical synchronization encodes multiple aspects of social interactions in monkey pairs.' *Sci Rep* 8, 4699 https://doi.org/10.1038/s41598-018-22679-x

Leong, Victoria *et al.* (2017) 'Speaker gaze increases information coupling between infant and adult brains.' *PNAS* 114 (50):13290-5

Critchlow, Hannah (2019) *The Science of Fate: The New Science of Who We Are – And How to Shape our Best Future.* London, UK: Hodder & Stoughton

Woolley, A.W. *et al.* (2010) 'Evidence for a collective intelligence factor in the performance of human groups.' *Science* 330, 686–8

Riedl, C. *et al.* (2022) 'Quantifying collective intelligence in human groups [published correction appears in *Proc Natl Acad Sci USA.*

119(19):e2204380119]. *Proc Natl Acad Sci USA.* 2021;118(21):e 2005737118. doi:10.1073/pnas.2005737118

Von Mohr, M. *et al.* (2022) 'Individuals with higher interoceptive accuracy are less suggestible to other people's judgements.' https://doi.org/10.31234/osf.io/d3wsf

Galvez-Pol, A. *et al.* (2022) 'People can identify the likely owner of heartbeats by looking at individuals' faces.' *Cortex.* 151:176–87. doi:10.1016/j.cortex.2022.03.003. Epub ahead of print. PMID: 35430451

7. The Hive Mind and Humanity's Dark Side

Hedges, Chris (2003) 'What Every Person Should Know About War.' *New York Times*, 6 July https://www.nytimes.com/2003/07/06/books/chapters/what-every-person-should-know-about-war.html

World Health Organization (2019) *Suicide in the World: Global Health Estimates* https://www.who.int/teams/mental-health-and-substance-use/suicide-data

Contagion: How Groups Get Infected with Bad Ideas

De Waal F. B. M., Preston, S. D. (2017) 'Mammalian empathy: behavioural manifestations and neural basis.' *Nat Rev Neurosci.* 18(8):498–509. doi:10.1038/nrn.2017.72. Epub 29 Jun 2017. PMID: 28655877

Neumann, R., Strack, F. (2000) '"Mood contagion": the automatic transfer of mood between persons.' *J Pers Soc Psychol.* 79(2):211–23. doi:10.1037//0022-3514.79.2.211. PMID: 10948975

Isabella, Giuliana and Carvalho, Hamilton C. (2016) 'Emotional Contagion and Socialization', in *Emotions, Technology, and Behaviors*.

(2008) 'Dynamic spread of happiness in a large social network: longitudinal analysis over 20 years in the Framingham Heart Study.' *BMJ*; 337: a2338 doi:https://doi.org/10.1136/bmj.a2338

Guillory, J., *et al.* (2011) 'Upset now? Emotion contagion in distributed groups.' *Proc ACM CHI Conf on Human Factors in Computing Systems* (Association for Computing Machinery, New York), 745–8

Kramer, Adam D. I., Guillory, Jamie E. and Hancock, Jeffrey T. (2014)

'Experimental evidence of massive-scale emotional contagion through social networks.' *PNAS* 111 (24) 8788–90; https://doi.org/10.1073/pnas.1320040111

Neumann, R. and Strack, F. (2000) '"Mood contagion": the automatic transfer of mood between persons.' *J Pers Soc Psychol.* 79(2):211–23. doi:10.1037//0022-3514.79.2.211. PMID: 10948975

Isabella, Giuliana and Carvalho, Hamilton C. (2016) 'Emotional Contagion and Socialization', in *Emotions, Technology, and Behaviors.*

(2008) 'Dynamic spread of happiness in a large social network: longitudinal analysis over 20 years in the Framingham Heart Study.' *BMJ*; 337: a2338 doi:https://doi.org/10.1136/bmj.a2338

Guillory, J., *et al.* (2011) 'Upset now? Emotion contagion in distributed groups.' *Proc ACM CHI Conf on Human Factors in Computing Systems* (Association for Computing Machinery, New York), 745–8

Kramer, Adam D. I., Guillory, Jamie E. and Hancock, Jeffrey T. (2014) 'Experimental evidence of massive-scale emotional contagion through social networks.' *PNAS* 111 (24) 8788–90; https://doi.org/10.1073/pnas.1320040111

Carnevale, P. J. and Isen, A. M. (1986) 'The influence of positive affect and visual access on the discovery of integrative solutions in bilateral negotiation.' *Organizational Behavior and Human Decision Processes,* 37(1), 1–13. https://doi.org/10.1016/0749-5978(86)90041-5

Barsade, Sigal G., Coutifaris, Constantinos G. V., Pillemer, Julianna (2018) 'Emotional contagion in organizational life.' Research in *Organizational Behavior,* vol. 38, 137–51

Iacoboni, Marco (2008) *Mirroring People: The New Science of How We Connect with Others* (first edition). Farrar, Straus and Giroux

Prochazkova, Eliska and Kret, Mariska E. (2017) 'Connecting minds and sharing emotions through mimicry: A neurocognitive model of emotional contagion.' *Neuroscience & Biobehavioral Reviews,* vol. 80, 99–114

Christov-Moore, Leonardo, Conway, Paul, Iacoboni, Marco (2017) 'Deontological Dilemma Response Tendencies and Sensorimotor Representations of Harm to Others.' *Frontiers in Integrative Neuroscience* doi:10.3389/fnint.2017.00034

Dimberg, Ulf, Thunberg, Monika, Elmehed, Kurt (2000) 'Unconscious

Facial Reactions to Emotional Facial Expressions.' Brief Report. Find in PubMed https://doi.org/10.1111/1467-9280.00221

Gino, Francesca, Ayal, Shahar, Ariely, Dan (2009) 'Contagion and Differentiation in Unethical Behavior: The Effect of One Bad Apple on the Barrel.' Research Article, *Psychological Science*, vol. 20, issue 3, 393–8 https://doi.org/10.1111/j.1467-9280.2009.02306.x

Barsade, Sigal G., Coutifaris, Constantinos G. V., Pillemer, Julianna (2018) 'Emotional contagion in organizational life.' Research in *Organizational Behavior*, vol. 38, 137–51

Friedman, Howard S. *et al.* (1980) 'Understanding and Assessing Nonverbal Expressiveness: The Affective Communication Test.' *Journal of Personality and Social Psychology*, vol. 39, no. 2, 333–51

Neumann, R. and Strack, F. (2000) '"Mood contagion": the automatic transfer of mood between persons.' *J Pers Soc Psychol.* 79(2):211–23. doi:10.1037//0022-3514.79.2.211. PMID: 10948975

Isabella, Giuliana and Carvalho, Hamilton C. (2016) 'Emotional Contagion and Socialization', in *Emotions, Technology, and Behaviors*.

(2008) 'Dynamic spread of happiness in a large social network: longitudinal analysis over 20 years in the Framingham Heart Study.' *BMJ*; 337: a2338 doi:https://doi.org/10.1136/bmj.a2338

Guillory J. *et al.* (2011) 'Upset now? Emotion contagion in distributed groups.' *Proc ACM CHI Conf on Human Factors in Computing Systems* (Association for Computing Machinery, New York), 745–8.

Kramer, Adam D. I., Guillory, Jamie E. and Hancock, Jeffrey T. (2014) 'Experimental evidence of massive-scale emotional contagion through social networks.' *PNAS* 111 (24) 8788–90; https://doi.org/10.1073/pnas.1320040111

Carnevale, P. J. and Isen, A. M. (1986) 'The influence of positive affect and visual access on the discovery of integrative solutions in bilateral negotiation.' *Organizational Behavior and Human Decision Processes*, 37(1), 1–13. https://doi.org/10.1016/0749-5978(86)90041-5

Barsade, Sigal G., Coutifaris, Constantinos G. V., Pillemer, Julianna (2018) 'Emotional contagion in organizational life.' Research in *Organizational Behavior*, vol. 38, 137–51.

Iacoboni, Marco (2008) *Mirroring People: The New Science of How We Connect with Others* (first edition). Farrar, Straus and Giroux.

Prochazkova, Eliska and Kret, Mariska E. (2017) 'Connecting minds and sharing emotions through mimicry: A neurocognitive model of emotional contagion.' *Neuroscience & Biobehavioral Reviews*, vol. 80, 99–114

Mirror neuron discovery: https://sitn.hms.harvard.edu/flash/2016/mirror-neurons-quarter-century-new-light-new-cracks

Di Pellegrino, G. *et al.* (1992) 'Understanding motor events: a neurophysiological study.' *Exp Brain Res* 91, 176–180. https://doi.org/10.1007/BF00230027

A caveat with mirror neurons: they may be lighting up simply from recording an action, nothing to do with anybody else's actions: Albertini D. *et al.* (2021) 'Largely shared neural codes for biological and nonbiological observed movements but not for executed actions in monkey premotor areas.' *J Neurophysiol*;126(3):906–12. doi:10.1152/jn.00296.2021. Epub 11 Aug 2021. PMID: 34379489.

A caveat with mirror neurons: they may be lighting up simply from recording an action, nothing to do with anybody else's actions:

Napolitano, Anna (2021) 'Study casts new light on mirror neurons.' nature.com, 24 August https://www.nature.com/articles/d43978-021-00101-x

Wingenbach, T. S. H. *et al.* (2020) 'Perception of Discrete Emotions in Others: Evidence for Distinct Facial Mimicry Patterns.' *Sci Rep* 10, 4692. https://doi.org/10.1038/s41598-020-61563-5

Di Pellegrino, G. *et al.* (1992) 'Understanding motor events: a neurophysiological study.' *Exp Brain Res* 91, 176–180. https://doi.org/10.1007/BF00230027

Gendron, M., Crivelli, C., Barrett, L. F. (2018) 'Universality Reconsidered: Diversity in Making Meaning of Facial Expressions.' *Curr Dir Psychol Sci.* 27(4):211–19. doi:10.1177/0963721417746794. Epub 31 Jul 2018. PMID: 30166776; PMCID: PMC6099968

Lockwood, P. L. *et al.* (2017) 'Individual differences in empathy are associated with apathy-motivation.' *Sci Rep* 7, 17293 https://doi.org/10.1038/s41598-017-17415-w

Gansberg, Martin (1964) '37 Who Saw Murder Didn't Call the Police; Apathy at Stabbing of Queens Woman Shocks Inspector.' *New York Times*, 27 March. New York, NY: *New York Times* Company

Hudson, James M. and Bruckman, Amy S. (2004) 'The Bystander

Effect: A Lens for Understanding Patterns of Participation.' *Journal of the Learning Sciences*. 13 (2): 165–95. CiteSeerX 10.1.1.72.4881. doi:10.1207/s15327809jls1302_2

Meyers, D. G. (2010) *Social Psychology* (tenth edition). New York: McGraw-Hill

Christensen, K. and Levinson, D. (2003) *Encyclopedia of community: From the village to the virtual world,* Band 1, 662

Power and Conformity: Flip-sides of the Same Coin

The Stanford Prison Experiment: https://www.prisonexp.org

Bekiempis, Victoria (2015) 'What Philip Zimbardo and the Stanford Prison Experiment Tell Us About Abuse of Power.' *Newsweek*, 4 August

Galinsky, A. D. *et al.* (2006) 'Power and perspectives not taken.' *Psychological Science*, 17, 1068–74

Gruenfeld, D. *et al.* (2008). 'Power and the objectification of social targets.' *Journal of Personality and Social Psychology*, 95, 111–27.

Hogeveen, J., Inzlicht, M., Obhi, S. S. (2014) 'Power changes how the brain responds to others.' *J Exp Psychol Gen.* 143(2):755–62. doi:10.1037/a0033477. Epub 1 Jul 2013. PMID: 23815455

Keltner, D., Gruenfeld, D. and Anderson, C. (2003) 'Power, approach, and inhibition.' *Psychological Review*, 110, 265–84

Cho, M., Keltner, D. (2020) 'Power, approach, and inhibition: empirical advances of a theory.' *Curr Opin Psychol.* 33:196–200. doi:10.1016/j.copsyc.2019.08.013. Epub 22 Aug 2019. PMID: 31563791

Kogan, A. *et al.* (2014) 'Vagal activity is quadratically related to prosocial traits, prosocial emotions, and observer perceptions of prosociality.' *J Pers Soc Psychol.* 107(6):1051–63. doi:10.1037/a0037509. Epub 22 Sep 2014. PMID: 25243414

Anderson, C., Berdahl, J. L. (2002) 'The experience of power: Examining the effects of power on approach and inhibition tendencies.' *Journal of Personality and Social Psychology*, 83, 1362–77

Youssef, F. F. *et al.* 'Sex differences in the effects of acute stress on behavior in the ultimatum game.' *Psychoneuroendocrinology*, 96, 126–31

Nowak, M. A. (2000) 'Fairness Versus Reason in the Ultimatum Game.' *Science*. 289 (5485): 1773–5. doi:10.1126/science.289.5485.1773. PMID:10976075

The Urge to Conform: A Risk Factor for Collective Stupidity

Blakemore, Sarah-Jayne (2018) *Inventing Ourselves: The Secret Life of the Teenage Brain* (first edition). Doubleday

Asch, Solomon (1956) 'Studies of independence and conformity: I. A minority of one against a unanimous majority.' *Psychological Monographs: General and Applied*. 70 (9): 1–70. doi:10.1037/h0093718

Asch, Solomon (1955) 'Opinions and social pressure.' Readings about the social animal. 17–26

Stout, D. (1996) 'Solomon Asch is dead at 88; a leading social psychologist.' *New York Times*. 29 February

Browning, Christopher R. (1998) [1992] *Ordinary Men: Reserve Police Battalion 101 and the Final Solution in Poland*. New York: Harper Perennial, 171ff

Caspar, E. A. *et al.* (2020) 'Obeying orders reduces vicarious brain activation towards victims' pain.' *NeuroImage*, 222, 117251. https://doi.org/10.1016/j.neuroimage.2020.117251

From Psychopaths to Hyper-altruists: Ethical Thinking

Masserman J. H., Wechkin, S., Terris, W. '"Altruistic" behavior in rhesus monkeys.' *Am J Psychiatry*. 1964;121:584–5.

Leeks, A., West, S. (2019) 'Altruism in a virus.' *Nat Microbiol* 4, 910–11 https://doi-org.ezp.lib.cam.ac.uk/10.1038/s41564-019-0463-0

Bourke, A. F. G. (2021) 'The role and rule of relatedness in altruism.' *Nature* 590(7846):392–4. doi:10.1038/d41586-021-00210-z. PMID: 33526901

Kay, T., Keller, L. and Lehmann, L. (2020) 'The evolution of altruism and the serial rediscovery of the role of relatedness.' *Proc Natl Acad Sci. USA* 117, 28894–8

Cesarini, D. *et al.* (2009) 'Genetic variation in preferences for giving and risk taking.' *Quart Econ* 124, 809–42. doi:10.1162/qjec.2009.124.2.809

Gregory, A. M. *et al.* (2009) 'Behavioral genetic analyses of prosocial behavior in adolescents.' *Dev Sci* 12, 165–74. (doi:10.1111/j.1467-7687.2008.00739.x

Hur, Y. M., Rushton, J. P. (2007) 'Genetic and environmental contributions to prosocial behaviour in 2-to 9-year-old South Korean twins.' *Biol Lett.* 3, 664–6. doi:10.1098/rsbl.2007.0365

Israel, S., Hasenfratz, L., Knafo-Noam, A. (2015) 'The genetics of morality and prosociality.' *Curr Opin Psychol.* 6, 55–9. doi:10.1016/j.copsyc.2015.03.027

Knafo, A., Plomin, R. (2006) 'Prosocial behavior from early to middle childhood: genetic and environmental influences on stability and change.' *Dev Psychol.* 42, 771–86. doi:10.1037/0012-1649.42.5.771

Wang, C., Lu, X. (2018) 'Hamilton's inclusive fitness maintains heritable altruism polymorphism through rb = c.' *Proc Natl Acad Sci USA* 115:1860–4

Sibly, Richard M. and Curnow, Robert N. (2017) 'Royal Society Open Science: Genetic polymorphisms between altruism and selfishness close to the Hamilton threshold rb = c.' https://doi org.ezp.lib.cam.ac.uk/10.1098/rsos.160649

Laursen, H. R. *et al.* (2014) 'Variation in the oxytocin receptor gene is associated with behavioural and neural correlates of empathic accuracy.' *Front Behav Neurosci*, 8, 423

Walter, N. T. (2012) 'Ignorance is no excuse: moral judgments are influenced by a genetic variation on the oxytocin receptor gene.' *Brain Cogn*, 78, 268–73

Marsh, A. A. *et al.* (2011) 'Serotonin transporter genotype (5-HTTLPR) predicts utilitarian moral judgments.' *PLoS One*, 6, e25148

Greenberg, D., Huppert, J. D. (2010) 'Scrupulosity: A Unique Subtype of Obsessive-Compulsive Disorder.' *Curr Psychiatry Rep* 12, 282–9. https://doi.org/10.1007/s11920-010-0127-5

Miller, C. H., Hedges, D. W. (2008) 'Scrupulosity disorder: an overview and introductory analysis.' *J Anxiety Disord.* 22(6):1042–58. doi:10.1016/j.janxdis.2007.11.004. Epub 21 Nov 2007. PMID: 18226490

Crockett, Molly J. (2014) 'Harm to others outweighs harm to self.' *Proceedings of the National Academy of Sciences*, 111 (48) 17320–5; doi:10.1073/pnas.1408988111

Crockett, M.J. (2017) 'Moral transgressions corrupt neural representations of value.' *Nat Neurosci.* 20(6):879–85. doi:10.1038/nn.4557. Epub 1 May 2017. PMID: 28459442; PMCID: PMC5462090

Carlson, R. W., Crockett, M. J. (2018) 'The lateral prefrontal cortex and moral goal pursuit.' *Curr Opin Psychol.* 24:77–82. doi:10.1016/j. copsyc.2018.09.007. Epub 1 Oct 2018. PMID: 30342428

Yu, H. *et al.* (2020) 'Toward a Brain-Based Bio-Marker of Guilt.' *Neurosci Insights.*15:2633105520957638.doi:10.1177/2633105520957638

Yu, H. *et al.* (2014) 'The voice of conscience: Neural bases of interpersonal guilt and compensation.' *Social Cognitive and Affective Neuroscience*, 9(8), 1150–8 (journal link)

Yu, H. *et al.* (2020). 'A generalizable multivariate brain pattern for interpersonal guilt.' *Cerebral Cortex*, 30(6), 3558–2. Pre-print: (journal link)

Nicolle, A. *et al.* (2011) 'A role for the striatum in regret-related choice repetition.' *J Cogn Neurosci.* 23(4):845–56. doi:10.1162/jocn.2010.21510

'Are Moral Values Contagious?' Interview with Professor Ray Dolan, University College, 6 July 2014. londonhttps://www.thenakedscientists. com/articles/interviews/are-moral-values-contagious

Seppälä, Emma M. (ed.) (2017) *The Oxford Handbook of Compassion Science* (Oxford Library of Psychology). OUP USA

Taber-Thomas, B. C. *et al.* (2014) 'Arrested development: early prefrontal lesions impair the maturation of moral judgement.' *Brain.* 137 (Pt. 4): 1254–61. doi:10.1093/brain/awt377. PMC 3959552. PMID:24519974

Bechara, A., Tranel, D., Damasio, H. (2000) 'Characterization of the decision-making deficit of patients with ventromedial prefrontal cortex lesions.' *Brain.* 123 (Pt. 11) (11): 2189–2202. Doi:10.1093/ brain/123.11.2189

Fan, Y. *et al.* (2011) 'Is there a core neural network in empathy? An fMRI based quantitative meta-analysis.' *Neurosci Biobehav Rev.* 35(3):903–11. doi:10.1016/j.neubiorev.2010.10.009. Epub 23 Oct 2010. PMID: 20974173

Lamm, C., Decety, J., Singer, T. (2011) 'Meta-analytic evidence for common and distinct neural networks associated with directly experienced pain and empathy for pain.' *Neuroimage.* 1;54(3):2492–502. doi:10.1016/j.neuroimage.2010.10.014. Epub 12 Oct 2010. PMID: 20946964

Darwin, C. (1871) *The Descent of Man and Selection in Relation to Sex (vol 1).* London, UK: Murray

Tomasello, M., Vaish, A. (2013) 'Origins of human cooperation and morality.' *Annu Rev Psychol.* 64:231–55. doi:10.1146/annurev-psych-113011-143812. Epub 12 Jul 2012. PMID: 22804772

Ashar, Yoni *et al.* (2016) 'Toward a Neuroscience of Compassion'. 10.1093/acprof:oso/9780199977925.003.0009

Dunn, E. W., Aknin, L. B., Norton, M. I. (2008) 'Spending money on others promotes happiness.' *Science.* 21;319(5870):1687–8. doi:10.1126/science.1150952. Erratum in: *Science.* 2009 May 29;324(5931):1143. PMID: 18356530

Ko, C. M. (2018) 'Effect of Seminar on Compassion on student self-compassion, mindfulness and well-being: A randomized controlled trial.' *J Am Coll Health.* 66(7):537–45. doi:10.1080/07448481.2018.1431913. Epub 22 Mar 2018. PMID: 29405863

Carson, J. W. *et al.* (2005) 'Loving-kindness meditation for chronic low back pain: results from a pilot trial.' *J Holist Nurs.* 23(3):287–304. doi:10.1177/0898010105277651. PMID: 16049118

Pace, T. W. *et al.* (2010) 'Innate immune, neuroendocrine and behavioral responses to psychosocial stress do not predict subsequent compassion meditation practice time.' *Psychoneuroendocrinology.* 35(2):310–15. doi:10.1016/j.psyneuen.2009.06.008. Epub 16 Jul. 2009. PMID: 19615827; PMCID: PMC3083925

Post, S. G. (2005) 'Altruism, happiness, and health: it's good to be good.' *Int J Behav Med.* 12(2):66–77

Hare, R. D., Neumann, C. S. (2008) 'Psychopathy as a clinical and empirical construct.' *Annu Rev Clin Psychol.* 4:217–46. doi:10.1146/annurev.clinpsy.3.022806.091452. PMID: 18370617

Cima, M., Tonnaer, F., Hauser. M. D. (2010) 'Psychopaths know right from wrong but don't care.' *Soc Cogn Affect Neurosci.* 5(1):59–67. doi:10.1093/scan/nsp051

https://www.telegraph.co.uk/news/2016/09/13/1-in-5-ceos-are-psychopaths-australian-study-finds

https://www.forbes.com/sites/jackmccullough/2019/12/09/the-psychopathic-ceo/?sh=314ff015791e

Babiak, Paul and Hare, Robert D. (2019) *Snakes in Suits: Understanding and Surviving the Psychopaths in Your Office* (revised edition). Harper Business

Landay, K., Harms, P. D., Credé, M. (2019) 'Shall we serve the dark lords? A meta-analytic review of psychopathy and leadership.' *Journal of Applied Psychology, [s.l.],* vol. 104, no. 1, *Leadership,* 183–96. doi:10.1037/apl0000357.Available from: https://search-ebscohost-com.ezp.lib.cam.ac.uk/login.aspx?direct=true&db=pdh&AN=2018-51219-001&site=ehost-live&scope=site. Accessed 22 July 2021

Murphy, J. M. (1976) 'Psychiatric labeling in cross-cultural perspective.' *Science.* 191(4231):1019–28

Fecteau, S., Pascual-Leone, A., Théoret, H. (2008) 'Psychopathy and the mirror neuron system: preliminary findings from a non-psychiatric sample.' *Psychiatry Res.* 160(2):137–44. doi:10.1016/j.psychres.2007.08.022. Epub 2 Jul 2008. PMID: 18599127

Motzkin, Julian C. *et al.* (2011) 'Reduced Prefrontal Connectivity in Psychopathy.' *Journal of Neuroscience* 31 (48): 17348–57 doi:10.1523/JNEUROSCI.4215-11.2011

Kiehl, Kent A. and Hoffman, Morris B. (2014) 'The Criminal Psychopath: History, Neuroscience, Treatment, and Economics.' Jurimetrics. Author manuscript; available in PMC. Published in final edited form as: *Jurimetrics.* 2011 Summer; 51: 355–97

Kiehl, Kent A. (2006) 'A cognitive neuroscience perspective on psychopathy: Evidence for paralimbic system dysfunction.' Psychiatry Res. 142(2–3): 107–128. Published in final edited form as Published online 19 May 2006. doi:10.1016/j.psychres.2005.09.013

Decety, Jean (2013) 'An fMRI study of affective perspective taking in individuals with psychopathy: imagining another in pain does not evoke empathy.' *Front Hum Neurosci.* https://doi.org/10.3389/fnhum.2013.00489

Hosking, Jay G. (2017) 'Disrupted Prefrontal Regulation of Striatal Subjective Value Signals in Psychopathy.' *Neuron,* vol. 95, issue 1, 221–31. E4. Open Archive doi:https://doi-org.ezp.lib.cam.ac.uk/10.1016/j.neuron.2017.06.030

Tiihonen, J. *et al.* (2020). 'Neurobiological roots of psychopathy.' *Mol Psychiatry* 25, 3432–41. https://doi-org.ezp.lib.cam.ac.uk/10.1038/s41380-019-0488-z

Rautiainen, M. R. *et al.* (2016) 'Genome-wide association study of antisocial personality disorder.' *Transl Psychiatry.* 6:e883

Recidivism of Prisoners Released in 30 States in 2005: Patterns from 2005 to 2010 – Update https://bjs.ojp.gov/library/publications/recidivism-prisoners-released-30-states-2005-patterns-2005-2010-update

Sterbenz, Christina (2014). 'Why Norway's prison system is so successful.' *Business Insider*, 11 December. Retrieved 17 June 2020.

Caldwell, Michael F., McCormick, David J., Umstead, Deborah (2007) 'Evidence of Treatment Progress and Therapeutic Outcomes Among Adolescents With Psychopathic Features.' *Criminal Justice and Behavior*, vol. 34 no. 5, 573–87. doi:10.1177/0093854806297511 http://citeseerx.ist.psu.edu/viewdoc/download?doi=10.1.1.981.5972&rep=rep1&type=pdf

Catmur, C., Walsh, V. and Heyes, C. (2007) 'Sensorimotor learning configures the human mirror system.' *Curr Biol.* 17, 1527–31

Chester, D. S. *et al.* (2016) 'Narcissism is associated with weakened frontostriatal connectivity: a DTI study.' *Soc Cogn Affect Neurosci.* 11(7):1036–40. doi:10.1093/scan/nsv069. Epub 5 Jun 2015. PMID: 26048178; PMCID: PMC4927024

Jauk, E., Kanske, P. (2021) 'Can neuroscience help to understand narcissism? A systematic review of an emerging field.' *Personal Neurosci.* 4:e3. doi:10.1017/pen.2021.1. PMID: 34124536; PMCID: PMC8170532

Paris, J. (2014) 'Modernity and narcissistic personality disorder.' *Personal Disord.* 5(2):220–6. doi:10.1037/a0028580. Epub 16 Jul 2012. PMID: 22800179

Twenge, J. M., Campbell, W. K. (2009) *The Narcissism Epidemic: Living in the Age of Entitlement.* New York, NY: Free Press

Newsom, C. R. *et al.* (2003) 'Changes in adolescent response patterns on the MMPI/MMPI-A across four decades.' *Journal of Personality Assessment.* 81(1):74–84. doi:10.1207/S15327752JPA8101_07. 99679-007

Twenge, J. M., Campbell, W. K., Gentile, B. (2013) 'Changes in pronoun use in American books and the rise of individualism, 1960–2008.' *Journal of Cross-Cultural Psychology.* 44(3):406–15. doi:10.1177/0022022112455100. 2013-06649-005.

Twenge, J. M., Campbell, W. K., Gentile, B. (2012) 'Increases in individualistic words and phrases in American books, 1960–2008.' *PLoS One.* 7(7):e40181 Epub 19 Jul 2019. doi:10.1371/journal. pone.0040181. PubMed Central PMCID: PMCPMC3393731

DeWall, C. N. *et al.* (2011) 'Tuning in to psychological change: Linguistic markers of psychological traits and emotions over time in popular U.S. song lyrics.' *Psychology of Aesthetics, Creativity, and the Arts.* 5(3):200–7. doi:10.1037/a0023195. 2011-05681-001

Uhls, Y., Greenfield, P. (2011) 'The rise of fame: An historical content analysis.' *Cyberpsychology: Journal of Psychosocial Research on Cyberspace.* 5: article 1

Twenge, J. M (2006) *Generation Me: Why Today's Young Americans are More Confident, Assertive, Entitled – And More Miserable Than Ever Before.* New York: Free Press (Simon and Schuster)

Twenge, J. M. (2013) 'Teaching generation me.' *Teaching of Psychology,* 40(1), 66.

Vater, Aline, Moritz, Steffen, Roepke, Stefan (2018) 'Does a narcissism epidemic exist in modern western societies? Comparing narcissism and self-esteem in East and West Germany.' *PLoS One.* 13(1): e0188287. Published online 24 Jan 2018. doi:10.1371/journal. pone.0188287

Greenfield, P. M. (2013) 'The changing psychology of culture from 1800 through 2000.' *Psychol Sci.* 24: 1722–31. PMID: 23925305

Zeng, R., Greenfield, P. M. (2015) 'Cultural evolution over the last 40 years in China: Using the Google Ngram viewer to study implications of social and political change for cultural values.' *Int J Psychol.* 50: 47–55. PMID: 25611928

Wheeler, M. A., McGrath, M. J., Haslam, N. (2019) 'Twentieth century morality: The rise and fall of moral concepts from 1900 to 2007.' *PLoS One* 14(2): e0212267. https://doi.org/10.1371/journal. pone.0212267

Walker, M. (2009). 'Enhancing genetic virtue: A project for twenty-first century humanity?' *Politics and the Life Sciences.* 28 (2): 27–47

Baccarini, E., Malatesti, L. (2017) 'The moral bioenhancement of psychopaths.' *J Med Ethics.* 43(10):697–701. doi:10.1136/ medethics-2016-103537. Epub 29 Mar 2017. PMID: 28356492

Crutchfield, P. (2019) 'Compulsory moral bioenhancement should be covert.' *Bioethics* 33(1):112–21. doi:10.1111/bioe.12496. Epub 29 Aug 2018. PMID: 30157295

https://www.vice.com/en/article/z3xw3x/new-research-vindicates-1972-mit-prediction-that-society-will-collapse-soon

https://www.theguardian.com/environment/earth-insight/2014/jun/04/scientists-limits-to-growth-vindicated-investment-transition-circular-economy

Persson, I., Savulescu, J. (2017) 'Moral Hard-Wiring and Moral Enhancement.' *Bioethics.* 31(4):286–95. doi:10.1111/bioe.12314. Epub 16 Mar 2017. PMID: 28300281; PMCID: PMC5639457

Persson, I. Savulescu, J. (2012a). *Unfit for the Future: The Need for Moral Enhancement.* New York, NY: Oxford University Press

Dubljević, V., Racine, E. (2017). 'Moral enhancement meets normative and empirical reality: Assessing the practical feasibility of moral enhancement neurotechnologies.' *Bioethics.* 31 (5):338–48. doi:10.1111/bioe.12355. PMID 28503833

Thunberg, Greta (2019) *No One Is Too Small to Make a Difference.* Bokus.com (in Swedish). Retrieved 22 June 2019

8. Reshaping Our World to Boost Our Collective Intelligence

Tweaking the Environment for Better Joined-up Thinking

Lowe, Robert J., Huebner, Gesche M., Oreszczyn, Tadj (2018) 'Possible future impacts of elevated levels of atmospheric CO_2 on human cognitive performance and on the design and operation of ventilation systems in buildings.' https://doi.org/10.1177/0143624418790129

Allen, Joseph G. *et al.* (2015) 'Associations of Cognitive Function Scores with Carbon Dioxide, Ventilation, and Volatile Organic Compound Exposures in Office Workers: A Controlled Exposure Study of Green and Conventional Office Environments.' *Environmental Health Perspectives.* 124 (6): 805–812. doi:10.1289/ehp.1510037. http://dx.doi.org/10.1289/ehp.1510037

Vehviläinen, Tommi *et al.* (2016) 'High indoor CO_2 concentrations in an office environment increases the transcutaneous CO_2 level and sleepiness during cognitive work.' *Journal of Occupational and Environmental Hygiene.* 13:1, 19–29, doi:10.1080/15459624.2015.107 6160

Chang, T. Y., Kajackaite, A. (2019) 'Battle for the thermostat: Gender and the effect of temperature on cognitive performance.' *PLoS One* 14(5): e0216362. https://doi.org/10.1371/journal.pone.0216362

Maths Problem Solution:

The bat has to cost $1 more than the ball.

$1.00 + $0.10 does equal $1.10 **BUT** if you take $1.00 − $0.10 you get $0.90.

So, the ball must cost $0.05, and the bat must cost $1.05 since $1.05 + $0.05 = $1.10

You can also use algebra to solve the problem:

$x + ($1.00 + x) = 1.10

$1.00 + 2x = 1.10

$2x = $1.10 − 1.00

$2x = 0.10

Finally, solve for x:

$x = 0.05

Check your work:

$x + ($1.00 + x) = 1.10, so

$0.05 + ($1.00 + $0.05) = 1.10

The Behavioural Insights Team/Nudge Unit: https://www.bi.team https://theconversation.com/male-vs-female-brains-having-a-mix-of-both-is-common-and-offers-big-advantages-new-research-153242

Medicating Away from Joined-up Thinking?

https://www.bbc.com/future/article/20200108-the-medications-that-change-who-we-are

https://theconversation.com/whats-the-point-of-paracetamol-66808

https://www.ageuk.org.uk/globalassets/age-uk/documents/reports-and-publications/reports-and-briefings/health--wellbeing/medication/190819_more_harm_than_good.pdf

Dewall, C.N. *et al.* (2010) 'Acetaminophen reduces social pain: behavioral and neural evidence.' *Psychol Sci.* 21(7):931–7. doi:10.1177/0956797610374741. Epub 14 Jun 2010. PMID: 20548058

Mischkowski, D., Crocker, J. and Way, B. M. (2019). 'A social analgesic? Acetaminophen (paracetamol) reduces positive empathy.' *Frontiers in Psychology*, 10, 538

Crockett, Molly J. *et al.* (2008) 'Serotonin Modulates Behavioral Reactions to Unfairness.' *Science.* 320 (5884): 1155577. Bibcode: 2008Sci...320.1739C. doi:10.1126/science.1155577. PMC 2504725. PMID:18535210

Kahane, G. *et al.* (2018) 'Moving on from the trolley dillemna: Beyond Sacrificial Harm: A Two-Dimensional Model of Utilitarian Psychology.' *Psychological Review*, 125(2), 131–64

Bolling, M.Y., Kohlenberg, R.J. (2004) 'Reasons for quitting serotonin reuptake inhibitor therapy: paradoxical psychological side effects and patient satisfaction.' *Psychother Psychosom.* 73(6):380–5. doi:10.1159/000080392. PMID: 15479994

Barnhart, W. J., Makela, E. H., Latocha, M. J. (2004) 'SSRI-induced apathy syndrome: a clinical review.' *J Psychiatr Pract.*10(3):196–9. doi:10.1097/00131746-200405000-00010. PMID: 15330228

Fava M. *et al.* (2006) 'A cross-sectional study of the prevalence of cognitive and physical symptoms during long-term antidepressant treatment.' *J Clin Psychiatry* 67(11):1754–9. doi:10.4088/jcp.v67n1113. PMID: 17196056

Goodwin, G. M. *et al.* (2017) 'Emotional blunting with antidepressant treatments: A survey among depressed patients.' *J Affect Disord.* 221:31–35. doi:10.1016/j.jad.2017.05.048. Epub 6 Jun 2017. PMID: 28628765.

https://www.theguardian.com/society/2018/aug/10/four-million-people-in-england-are-long-term-users-of-antidepressants

Brody, Debra J. M.P.H. and Gu, Qiuping M.D., Ph.D. (2020) 'Antidepressant Use Among Adults: United States, 2015–2018 National Centre for Health Statistics Data Brief No. 377.'

'Antidepressant use in England soars as pandemic cuts counselling access Exclusive: more than 6m people receive drugs as experts warn of Covid pandemic's effects on mental health.' https://www.theguardian.com/society/2021/jan/01/covid-antidepressant-use-at-all-time-high-as-access-to-counselling-in-england-plunges

Gansberg, Martin (1964) '37 Who Saw Murder Didn't Call the Police; Apathy at Stabbing of Queens Woman Shocks Inspector.' *New York Times*, 27 March. New York, NY: *New York Times* Company

Jobling, S. *et al.* (2006) 'Predicted exposures to steroid estrogens in U.K. rivers correlate with widespread sexual disruption in wild fish populations.' *Environmental Health Perspectives*, 114 (S-1), 32–9. ISSN: 0091-6765

Hamilton, P. B. *et al.* (2014) 'Populations of a cyprinid fish are self-sustaining despite widespread feminization of males.' *BMC Biology*, 12 (1), 1. ISSN: 1741-7007

Brodin, T. *et al.* (2013) 'Dilute concentrations of a psychiatric drug alter behavior of fish from natural populations.' *Science*. 339(6121):814–5. doi:10.1126/science.1226850. PMID: 23413353

Sundin J. *et al.* (2019) 'Behavioural alterations induced by the anxiolytic pollutant oxazepam are reversible after depuration in a freshwater fish.' *Sci Total Environ*. 665:390–9. doi:10.1016/j.scitotenv.2019.02.049. Epub 5 Feb 2019. PMID: 30772569

Cerveny, D. *et al.* (2020). 'Bioconcentration and behavioral effects of four benzodiazepines and their environmentally relevant mixture in wild fish.' *Science of the Total Environment*. 702, 134780

McCallum, E, *et al.* (2019) 'Investigating tissue bioconcentration and the behavioural effects of two pharmaceutical pollutants on sea trout (Salmo trutta) in the laboratory and field.' *Aquatic Toxicology* 207: 170–8

'Drugs in the Water: Harvard Medical School Open Letter' 1 June 2011 https://www.health.harvard.edu/newsletter_article/drugs-in-the-water

Hellström G. *et al.* (2016) 'GABAergic anxiolytic drug in water increases migration behaviour in salmon.' *Nat Commun.* 7:13460. doi:10.1038/ncomms13460. PMID: 27922016; PMCID: PMC5155400

Richmond, E. K. *et al.* (2018) 'A diverse suite of pharmaceuticals contaminates stream and riparian food webs.' *Nat Commun.*9(1):4491. doi:10.1038/s41467-018-06822-w. PMID: 30401828; PMCID: PMC6219508

Our Brains Online – the Good, the Bad and the Ugly

https://www.theguardian.com/technology/2020/feb/01/amy-orben-psychology-smartphones-affecting-brain-social-media-teenagers-mental-health

Vuorre, M., Orben, A., Przybylski, A.K. (in press) 'There is no evidence that associations between adolescents digital technology engagement and mental health problems have increased.' *Clinical Psychological Science.*

Orben, A., Weinstein, N. and Przybylski, A. K. (2020) 'Only holistic and iterative change will fix digital technology research.' *Psychological Inquiry.* Open-Access Version

Wegner, D. M., Erber, R., Raymond, P. (1991) 'Transactive memory in close relationships.' *J Pers Soc Psychol.* 61(6):923–9. doi:10.1037//0022-3514.61.6.923. PMID: 1774630

Wegner, D. M. (1987) 'Transactive Memory: A Contemporary Analysis of the Group Mind.' In: Mullen, B., Goethals, G. R. (eds) *Theories of Group Behavior* (Springer Series in Social Psychology), New York, USA: Springer

Sparrow, B., Liu, J., Wegner, D. M. (2011) 'Google effects on memory: cognitive consequences of having information at our fingertips.' *Science.* 333(6043):776–8. doi:10.1126/science.1207745. Epub 14 Jul 2011. PMID: 21764755

Hollingshead, A. B. (1998) 'Retrieval processes in transactive memory systems.' *Journal of Personality and Social Psychology,* 74(3), 659–71. https://doi.org/10.1037/0022-3514.74.3.659

Hollingshead, A. (2001) 'Cognitive interdependence and convergent expectations in transactive memory.' *Journal of Personality and Social Psychology.* 81 (6): 1080–9. doi:10.1037/0022-3514.81.6.1080. PMID: 11761309

Hewitt, L.Y., Roberts, L. D. (2015) 'Transactive memory systems scale for couples: development and validation.' *Front Psychol*.6:516. doi:10.3389/fpsyg.2015.00516. PMID: 25999873; PMCID: PMC4419599

Harris, Celia B. *et al.*(2021) 'It's not who you lose, it's who you are: Identity and symptom trajectory in prolonged grief.' *Current Psychology* (New Brunswick, N.J.), 1–11. doi:10.1007/s12144-021-02343-w

Hinsz, V. B.; Tindale, R. S.; Vollrath, D. A. (1997) 'The emerging conceptualization of groups as information processors.' *Psychological Bulletin*. 121 (1): 43–64. doi:10.1037/0033-2909.121.1.43. PMID: 9000891

Liang, D.W.; Moreland, R. L.; Argote, L. (1995) 'Group versus individual training and group performance: The mediating role of transactive memory.' *Personality and Social Psychology Bulletin*. 21 (4): 384–93. doi:10.1177/0146167295214009

Xiongfei, Cao and Ahsan, Alib (2018) 'Enhancing team creative performance through social media and transactive memory system.' *International Journal of Information Management*, vol. 39, 69–79

Fear: Too Much Bad News Makes Us Stupid

https://amara.org/en/videos/ajUSZC5DgugU/info/dear-facebook-this-is-how-youre-breaking-democracy/

https://www.wsj.com/articles/facebook-knows-it-encourages-division-top-executives-nixed-solutions-11590507499?mod=hp_lead_pos5

APA Stress in America™ Survey: US at 'Lowest Point We Can Remember' (2017) Future of Nation Most Commonly Reported Source of Stress https://www.apa.org/news/press/releases/2017/11/lowest-point

Block, P. and Burnett Heyes, S. (2020) 'Sharing the Load: Contagion and tolerance of mood in social networks.' *Emotion*.Advance online publication. https://doi.org/10.1037/emo0000952

The Mind and Mental Health: How Stress Affects the Brain: https://www.tuw.edu/health/how-stress-affects-the-brain/#:~:text=It%20can%20disrupt%20synapse%20regulation,responsible%20for%20memory%20and%20learning

Roiser J. P., Sahakian, B. J. (2013). 'Hot and cold cognition in depression.'

CNS Spectr. 18 (3): 139–49. doi:10.1017/S1092852913000072. PMID:23481353. S2CID: 34123889

Nord, C. L. *et al.* (2020) 'The neural basis of hot and cold cognition in depressed patients, unaffected relatives, and low-risk healthy controls: An fMRI investigation.' *Journal of Affective Disorders* vol. 274, 389–98. doi:10.1016/j.jad.2020.05.022

Seppälä, Emma M. (ed.) (2017) *The Oxford Handbook of Compassion Science* (Oxford Library of Psychology). OUP USA.

Fan, Y. *et al.* (2011) 'Is there a core neural network in empathy? An fMRI based quantitative meta-analysis.' *Neurosci Biobehav Rev.* 35(3):903–11. doi:10.1016/j.neubiorev.2010.10.009. Epub 23 Oct 2010. PMID: 20974173.

Lamm, C., Decety, J., Singer, T. (2011) 'Meta-analytic evidence for common and distinct neural networks associated with directly experienced pain and empathy for pain.' *Neuroimage.* 1;54(3):2492–502. doi:10.1016/j.neuroimage.2010.10.014. Epub 12 Oct 2010. PMID: 20946964

Lockwood, P. L. *et al.* (2017). 'Individual differences in empathy are associated with apathy-motivation.' *Sci Rep* 7, 17293 https://doi.org/10.1038/s41598-017-17415-w

'Australia passes social media law penalising platforms for violent content.' https://www.theguardian.com/media/2019/apr/04/australia-passes-social-media-law-penalising-platforms-for-violent-content

Polarisation and Hostility: How We Destroyed Debate

Jamie Bartlett on Twitter: https://twitter.com/jamiejbartlett/status/1175074115457359874

Hagey, Keach and Horwitz, Jeff (2021) 'Facebook Tried to Make Its Platform a Healthier Place. It Got Angrier Instead. Internal memos show how a big 2018 change rewarded outrage and that CEO Mark Zuckerberg resisted proposed fixes.' 15 Sept. https://www.wsj.com/articles/facebook-algorithm-change-zuckerberg-11631654215

Sydney, Emily (2019) *Disrespectful Democracy: The Psychology of Political Incivility.* Columbia University Press

The Facebook Files: A *Wall Street Journal* Investigation, 5 Oct 2021 https://www.wsj.com/livecoverage/facebook-whistleblower-frances-haugen-senate-hearing/card/AxUJoSioqe4Px8YzsGuc

https://www.theguardian.com/education/2019/mar/20/cambridge-university-rescinds-jordan-peterson-invitation

https://www.thetimes.co.uk/article/jordan-peterson-anti-pc-scholar-dropped-by-cambridge-over-islamophobia-shirt-msgzrqsw9

Cambridge University votes to safeguard free speech https://www.bbc.com/news/education-55246793

Spring, Victoria L., Cameron, Daryl C., Cikara, Mina (2018) 'The Upside of Outrage.' *Trends in Cognitive Sciences*, vol. 22, issue 12, 1067–9. https://doi.org/10.1016/j.tics.2018.09.006

Brady, W. J. and Crockett, M. J. (2019) 'How effective is online outrage?' *Trends in Cognitive Sciences*, 23(2), 79–80. https://static1.squarespace.com/static/538ca3ade4b090f9ef331978/t/5c64477c9140b7af8196affd/1550075772677/How+Effective+Is+Online+Outrage%3F.pdf

Crockett, M. J. (2017). 'Moral outrage in the digital age.' *Nat Hum Behav* 1, 769–71 https://doi.org/10.1038/s41562-017-0213-3

Salerno, Jessica M., Peter-Hagene, Liana C. (2013) 'The Interactive Effect of Anger and Disgust on Moral Outrage and Judgments.' *Psychological Science* vol. 24 issue 10, 2069–78 https://doi.org/10.1177/0956797613486988

Propaganda and Fake News: Weaponising Information

Lewandowsky, S. and Cook, J. (2020) *The Conspiracy Theory Handbook*. Available at http://sks.to/conspiracy

'Coronavirus, "Plandemic" and the seven traits of conspiratorial thinking.' 15 May 2020: https://theconversation.com/coronavirus-plandemic-and-the-seven-traits-of-conspiratorial-thinking-138483

Rollwage, M. *et al.* (2020) 'Confidence drives a neural confirmation bias.' *Nature Communications*. 11: 2634. PMID:32457308 doi:10.1038/s41467-020-16278-6

Schwartenbeck, P., FitzGerald, T. H. B. and Dolan, R. (2016) 'Neural signals encoding shifts in beliefs.' Neuroimage, 125, 578–86. doi:10.1016/j.neuroimage.2015.10.067

Sharot, T. *et al.* (2012) 'Selectively altering belief formation in the human brain.' *Proc Natl Acad Sci USA*. 109(42):17058–62. doi:10.1073/pnas.1205828109. Epub 24 Sep 2012

Hart, Joshua and Graether, Molly (2018) 'Something's Going on Here: Psychological Predictors of Belief in Conspiracy Theories.' *Journal of Individual Differences* doi:10.1027/1614-0001/a000268

https://theconversation.com/to-combat-conspiracy-theories-teach-critical-thinking-and-community-values-147314

Roozenbeek, Jon (2020) 'Susceptibility to misinformation about COVID-19 around the world.' *Royal Society Open Science* vol. 7, issue 10 https://doi.org/10.1098/rsos.201199

Kindness Is Our Best Defence

Saleh, N. *et al.* (2021) 'Active inoculation boosts attitudinal resistance against extremist persuasion techniques – A novel approach towards the prevention of violent extremism.' *Behavioural Public Policy*

Van der Linden, Sander (2020) 'The Paranoid Style in American Politics Revisited: An Ideological Asymmetry in Conspiratorial Thinking' *Political Psychology*, vol. 42, issue 1 February 2021, 23–51 https://doi.org/10.1111/pops.12681

Blastland, M. *et al.* (2020) 'Five rules for evidence communication.' *Nature* (587), 362–4

Van Steenbergen, H. *et al.* (2021) 'How positive affect buffers stress responses.' *Current Opinion in Behavioral Sciences* 39: 153160

'The Mind and Mental Health: How Stress Affects the Brain.' https://www.tuw.edu/health/how-stress-affects-the-brain/#:~:text=It%20can%20disrupt%20synapse%20regulation,responsible%20for%20memory%20and%20learning

Roiser J. P., Sahakian, B. J. (2013). 'Hot and cold cognition in depression.' *CNS Spectr.* 18 (3): 139–49. doi:10.1017/S1092852913000072. PMID:23481353. S2CID: 34123889

Tomasello, M., Vaish, A. (2013) 'Origins of human cooperation and morality.' *Annu Rev Psychol.* 64:231–55. doi:10.1146/annurev-psych-113011-143812. Epub 12 Jul 2012. PMID: 22804772

Ashar, Yoni *et al.* (2016) *Toward a Neuroscience of Compassion.* 10.1093/acprof:oso/9780199977925.003.0009

Dunn, E. W., Aknin, L. B., Norton, M. I. (2008) 'Spending money on

others promotes happiness.' *Science*. 319(5870):1687–8. doi:10.1126/
science.1150952. Erratum in: *Science*. 2009 May 29;324(5931):1143.
PMID: 18356530

Ko, C. M. *et al.* (2018) 'Effect of Seminar on Compassion on student
self-compassion, mindfulness and well-being: A randomized
controlled trial.' *J Am Coll Health*. 66(7):537–45. doi:10.1080/07448
481.2018.1431913. Epub 22 Mar 2018. PMID: 29405863

Seppälä, Emma M. (ed.) (2017) *The Oxford Handbook of Compassion
Science* (Oxford Library of Psychology). OUP USA

Kappes, A. *et al.* (2018) 'Concern for Others Leads to Vicarious
Optimism.' *Psychological Science*, 29(3), 379–89. https://static1.
squarespace.com/static/538ca3ade4b090f9ef331978/t5a749f3fe4966
b8870534750/1517592384127/2018_VicariousOptimism_Kappes.pdf

9. Cathedral Thinking: Healing From the Past and Building the Future

Krznaric, Roman (2020) *The Good Ancestor: How to Think Long Term
in a Short-Term World*: W. H. Allen

The Darwin Tree of Life Project: https://www.darwintreeoflife.org

Brancaccio, M. *et al.* (2019) 'Cell-autonomous clock of astrocytes drives
circadian behavior in mammals.' *Science* 363: 187–92

Hastings, M. H., Maywood, E. S., Brancaccio, M. (2018) 'Generation
of circadian rhythms in the suprachiasmatic nucleus.' *Nat Rev
Neurosci*. 19(8):453–69

Critchlow, Hannah (2018) *Consciousness: A Ladybird Expert Book* (The
Ladybird Expert Series 29), London, UK: Michael Joseph

LeDoux, Joe (2019) *The Deep History of Ourselves: The Four-Billion-Year
Story of How We Got Conscious Brains*: Viking

'What's the use of consciousness? How the stab of conscience made
us really conscious.' (2016) In *Where's The Action? The Pragmatic Turn
in Cognitive Science*, Engel, Andreas K., Friston, Karl and Kragic,
Danica (eds) Cambridge, Mas: MIT Press doi:10.7551/mitp
ress/9780262034326.003.0012

Clausi, S. *et al.* (2015) 'Cerebellar damage impairs the self-rating of regret
feeling in a gambling task.' *Front Behav Neurosci*. 9:113. doi:10.3389/
fnbeh.2015.00113. PMID: 25999829; PMCID: PMC4419712

Halton, Mary (2019) 'If you want to take on big problems, try thinking like a bee.' 1 Jan / https://ideas.ted.com/if-you-want-to-tackle-big-problems-try-thinking-like-a-bee/

Future-focused Thinking: Building the Future We Need

BBC Radio 4: 'Will humans survive the century?' 11 March 2019 https://www.cser.ac.uk/news/bbc-radio-4-will-humans-survive-century

Sacks, Jonathan *Morality: Restoring the Common Good in Divided Times.* London, UK: Hodder & Stoughton

'Environmental activist Greta Thunberg calls for "cathedral thinking" to climate change.' *Irish Times*, 17 Apr 2019

https://www.irishtimes.com/news/environment/environmental-activist-greta-thunberg-calls-for-cathedral-thinking-to-climate-change-1.3863358

The Darwin Tree of Life Project: https://www.darwintreeoflife.org

BBC Radio 4: *The Cathedral Thinkers*. 30 March 2020 https://www.bbc.co.uk/programmes/m000gl8n

All Party Parliamentary Group for Future Generations: https://www.appgfuturegenerations.com

Lord Martin Rees quote, Intergeneration Foundation: https://www.if.org.uk/quote

'Cambridge Students Join Forces with MPs to Launch APPG Combating Political Short-Termism.' 22 January 2018 https://www.cser.ac.uk/news/cambridge-students-join-forces-mps-launch-appg-com

Krznaric, Roman (2020) *The Good Ancestor: How to Think Long Term in a Short-Term World*: W. H. Allen

'Here come the Time Rebels! Japan's "Future Design" movement shows how to factor future generations into our politics.' 25 October 2020 https://www.thealternative.org.uk/dailyalternative/2020/10/25/future-design-japan-time-rebels

Lessons From Our Ancestors

Yunkaporta, Tyson (2020) *Sand Talk: How Indigenous Thinking Can Save the World*: HarperOne

Ubuntu: I am because we are https://olivenetwork.org/Issue/ubuntu-i-am-because-we are/24347#:~:text=An%20anthropologist%20proposed%20a%20game,sat%20together%20enjoying%20their%20treats

Rakoff, V. (1966) 'A long term effect of the concentration camp experience.' *Viewpoints* 1:17–22

Yehuda, R., Lehrner, A. (2018) 'Intergenerational transmission of trauma effects: Putative role of epigenetic mechanisms.' *World Psychiatry* 17:243–257

Bierer, L. M. *et al.* (2020) 'Intergenerational effects of maternal holocaust exposure on FKBP5 methylation.' *American Journal of Psychiatry* vol. 177, issue: 8, 744–53. ISSN: 1535-7228

Costa, Dora L., Yetter, Noelle and DeSomer, Heather (2018) 'Intergenerational transmission of paternal trauma among US Civil War ex-POWs.' *PNAS* 115 (44) 11215–20; first published 15 October 2018 https://doi.org/10.1073/pnas.1803630115

Curry, A. (2019) 'A painful legacy.' *Science*. 365(6450):212–15. doi:10.1126/science.365.6450.212. PMID: 31320518

Gillson, S. L., Ross, D. A. (2019) 'From Generation to Generation: Rethinking "Soul Wounds" and Historical Trauma.' *Biol Psychiatry*. 86(7):e19-e20. doi:10.1016/j.biopsych.2019.07.033. PMID: 31521209; PMCID: PMC7557912

Embodied Experience: Turning Trauma Into Resilience

Felitti, V. J. *et al.* (1998) 'Relationship of childhood abuse and household dysfunction to many of the leading causes of death in adults: The Adverse Childhood Experiences (ACE) Study.' *Am J Prev Med.* 14(4):245–58. doi:10.1016/s0749-3797(98)00017-8. PMID: 9635069

Felitti, V. J. (2002) 'The Relation Between Adverse Childhood Experiences and Adult Health: Turning Gold into Lead.' *Perm J*.6(1):44–7. PMID: 30313011; PMCID: PMC6220625

Kezelman, C. *et al.* (2015) 'The Cost of Unresolved Childhood Trauma and Abuse in Adults in Australia, Adults Surviving Child Abuse and Pegasus Economics, Sydney

McCarthy, M. M. *et al.* (2016) 'The lifetime economic and social costs of child maltreatment in Australia.' *Children and Youth Services Review*, 71, 217–226

Belsky, Jay (2020) *The Origins of You: How Childhood Shapes Later Life*.

https://www.sciencemag.org/news/2018/02/two-psychologists-followed-1000-new-zealanders-decades-here-s-what-they-found-about-how

Anne-Laura van Harmelen, Professor of Brain, Safety and Resilience https://www.universiteitleiden.nl/en/staffmembers/anne-laura-van-harmelen#tab-1

Thomason, M. E. and Marusak, H. A. (2017) 'Toward understanding the impact of trauma on the early developing human brain.' *Neuroscience*.342:55–67. doi:10.1016/j.neuroscience.2016.02.022. Epub 15 Feb 2016. PMID: 26892294; PMCID: PMC4985495

Cacioppo, John T. *et al.* (eds.) (2002) *Foundations in Social Neuroscience*. Cambridge, Mass.: MIT Press

Tyborowska, Anna *et al.* (2018) 'Early-life and pubertal stress differentially modulate grey matter development in human adolescents.' *Scientific Reports*. doi:10.1038/s41598-018-27439-5

Moreno-Lopez, Laura (2021) 'Early adolescent friendships aid behavioural and neural responses to social exclusion in young adults.' https://psyarxiv.com/zfh5m

Roth, T. L. *et al.* (2009) 'Lasting epigenetic influence of early-life adversity on the BDNF gene.' *Biol Psychiatry* 65(9):760–9. doi:10.1016/j.biopsych.2008.11.028. Epub 15 Jan 2009. PMID: 19150054; PMCID: PMC3056389

Ganzel, B. *et al.* (2007) 'The aftermath of 9/11: effect of intensity and recency of trauma on outcome.' *Emotion*.7(2):227–38. doi:10.1037/1528-3542.7.2.227

Hamwey, Meghan K. *et al.* (2020) 'Post-Traumatic Stress Disorder among Survivors of the September 11, 2001 World Trade Center Attacks: A Review of the Literature.' *International Journal of Environmental Research and Public Health* vol. 17,12 4344. doi:10.3390/ijerph17124344

Galea S. *et al.* (2002) 'Posttraumatic stress disorder in Manhattan, New York City, after the September 11th terrorist attacks.' *J. Urban Health*. 79:340–53. doi:10.1093/jurban/79.3.340

The Legacy of Trauma and Resilience – How Knowledge Is Passed On

Rakoff, V. (1966): 'A long term effect of the concentration camp experience.' *Viewpoints* 1:17–22

Gillson, S. L., Ross, D. A. (2019) 'From Generation to Generation: Rethinking "Soul Wounds" and Historical Trauma.' *Biol Psychiatry.* 86(7):e19-e20. doi:10.1016/j.biopsych.2019.07.033. PMID: 31521209; PMCID: PMC7557912

Dias, B. G., Ressler, K. J. (2014) 'Parental olfactory experience influences behavior and neural structure in subsequent generations.' *Nat Neurosci.* 17(1):89–96. doi:10.1038/nn.3594. Epub 1 Dec 2013. PMID: 24292232; PMCID: PMC3923835

Aoued, H. S. *et al.* (2019) 'Reversing Behavioral, Neuroanatomical, and Germline Influences of Intergenerational Stress.' *Biol Psychiatry.* 85(3):248–56. doi:10.1016/j.biopsych.2018.07.028. Epub 27 Aug 2018. PMID: 30292395; PMCID: PMC6326876

Yehuda, Rachel *et al.* (2015) 'Holocaust Exposure Induced Intergenerational Effects on *FKBP5* Methylation.' *Biological Psychiatry*, Archival Report vol. 80, issue 5, 372–80 doi:https://doi.org/10.1016/j.biopsych.2015.08.005

Dias, B. G., Ressler, K. J. (2014) 'Parental olfactory experience influences behavior and neural structure in subsequent generations.' *Nat Neurosci.* 17(1):89–96. doi:10.1038/nn.3594. Epub 1 Dec 2013. PMID: 24292232; PMCID: PMC3923835

Aoued, H. S. *et al.* (2019) 'Reversing Behavioral, Neuroanatomical, and Germline Influences of Intergenerational Stress.' *Biol Psychiatry.* 85(3):248–56. doi:10.1016/j.biopsych.2018.07.028. Epub 2018 Aug 27. PMID: 30292395; PMCID: PMC6326876

Costa, Dora L., Yetter, Noelle and DeSomer, Heather (2018) 'Intergenerational transmission of paternal trauma among US Civil War ex-POWs.' *PNAS* 115 (44) 11215–20; first published 15 October 2018 https://doi.org/10.1073/pnas.1803630115

Curry, A. (2019) 'A painful legacy.' *Science.* 365(6450):212–15. doi:10.1126/science.365.6450.212. PMID: 31320518

Gillson, S. L., Ross, D. A. (2019) 'From Generation to Generation: Rethinking "Soul Wounds" and Historical Trauma.' *Biol Psychiatry.*

86(7):e19-20. doi:10.1016/j.biopsych.2019.07.033. PMID: 31521209; PMCID: PMC7557912

Lamarck, J. B. (1809) *Philosophie zoologique*. Paris: Dentu et L'Auteur

Darwin, Charles (1859) *On the Origin of Species*: London, John Murray.

Darwin, C. (1873) 'Inherited Instinct.' *Nature* 7, 281 https://doi.org/10.1038/007281b0

Darwin, C. (1871) 'Pangenesis.' *Nature* 3, 5023 https://doi.org/10.1038/003502a0

Liu, Y., Chen, Q. (2018) '150 years of Darwin's theory of intercellular flow of hereditary information.' *Nat Rev Mol Cell Biol* 19, 749–50. https://doi.org/10.1038/s41580-018-0072-4

Liu, Y. (2019) 'Darwin and *Nature's* 150th anniversary.' *Nature*. 574(7776):36. doi:10.1038/d41586-019-02927-4. PMID: 31576045

Szyf, M. (2014) 'Lamarck revisited: epigenetic inheritance of ancestral odor fear conditioning.' *Nat Neurosci*. 17(1):2–4. doi:10.1038/nn.3603. PMID: 24369368

Welberg, L. (2014) 'Epigenetics: a lingering smell?' *Nat Rev Neurosci*. 15(1):1. doi:10.1038/nrn3660. PMID: 24356065

Ortela, A. and Esteller, M. (2010) 'Epigenetic modifications and human disease.' *Nature Biotechnology*, 28, 1057–68

Rivera, R. M. and Bennett, L. B. (2010) 'Epigenetics in humans: An overview.' *Current Opinion in Endocrinology, Diabetes and Obesity*, 17, 493–9

Yehuda, Rachel *et al*. (2015) 'Holocaust Exposure Induced Intergenerational Effects on *FKBP5* Methylation.' *Biological Psychiatry*, Archival Report vol. 80, issue 5, 372–80 doi:https://doi.org/10.1016/j.biopsych.2015.08.005

Daskalakis, N. P. *et al*. (2021) 'Intergenerational trauma is associated with expression alterations in glucocorticoid- and immune-related genes.' *Neuropsychopharmacol*. 46, 763–73 https://doi.org/10.1038/s41386-020-00900-8

Bierer, L. M. *et al*. (2020) 'Intergenerational Effects of Maternal Holocaust Exposure on *FKBP5* Methylation.' *Am J Psychiatry*. 177(8):744–53. doi:10.1176/appi.ajp.2019.19060618. Epub 21 Apr 2020. PMID: 32312110

Daskalakis, N. P. *et al*. (2021) 'Intergenerational trauma is associated with expression alterations in glucocorticoid- and immune-related genes.' *Neuropsychopharmacol.* 46, 763–73 https://doi.org/10.1038/s41386-020-00900-8

Klosin, A. *et al.* (2017) 'Transgenerational transmission of environmental information in *C. elegans.' Science.* 356(6335):320–3. doi:10.1126/science.aah6412. PMID: 28428426

Creating Growth Out of Pain

Kaldewaij, R. *et al.* (2021). 'Anterior prefrontal brain activity during emotion control predicts resilience to post-traumatic stress symptoms.' *Nat Hum Behav* 5, 1055–64. https://doi.org/10.1038/s41562-021-01055-2

Bramson, B. *et al.* (2020). 'Improving emotional-action control by targeting long-range phase-amplitude neuronal coupling.' *eLife.* 9:e59600. Full text https://elifesciences.org/articles/59600

Michela, A. *et al.* (2022) 'Deep-Breathing Biofeedback Trainability in a Virtual-Reality Action Game: A Single-Case Design Study With Police Trainers.' *Front Psychol.* 13:806163. doi:10.3389/fpsyg.2022.806163. PMID: 35222194; PMCID: PMC8868154

Tyborowska, Anna *et al.* (2018) 'Early-life and pubertal stress differentially modulate grey matter development in human adolescents.' *Scientific Reports* doi:10.1038/s41598-018-27439-5

Koch, S. B. J. et al. (2021) 'Larger dentate gyrus volume as predisposing resilience factor for the development of trauma-related symptoms.' *Neuropsychopharmacol.* https://doi.org/10.1038/s41386-020-00947-7

Van Praag, H., Kempermann, G. and Gage, F. (1999) 'Running increases cell proliferation and neurogenesis in the adult mouse dentate gyrus.' *Nat Neurosci* 2, 266–70. https://doi.org/10.1038/6368

Nauer, R. K. *et al.* (2020) 'Improving fitness increases dentate gyrus/CA3 volume in the hippocampal head and enhances memory in young adults.' *Hippocampus.* 30(5):488–504. doi:10.1002/hipo.23166. Epub 7 Oct 2019. PMID: 31588607; PMCID: PMC7485880

Kaldewaij, R. *et al.* (2021) 'Anterior prefrontal brain activity during emotion control predicts resilience to post-traumatic stress symptoms.' *Nat Hum Behav* https://doi.org/10.1038/s41562-021-01055-2

Hoffman, Benjamin U., Lumpkin, Ellen A. (2018) 'A gut feeling.' *Science* vol. 361, issue 6408, 1203–4 doi:10.1126/science.aau9973

Le Chatelier, E. *et al.* (2013). 'Richness of human gut microbiome correlates with metabolic markers.' *Nature* 500, 541–6 https://doi.org/10.1038/nature12506

Mims, T. S. *et al.* (2021) 'The gut mycobiome of healthy mice is shaped by the environment and correlates with metabolic outcomes in response to diet.' *Commun Biol* 4, 281 https://doi.org/10.1038/s42003-021-01820-z

Riquelme, Erick *et al.* (2019) 'Tumor Microbiome Diversity and Composition Influence Pancreatic Cancer Outcomes.' *Cell* 178(4):795–806.e12. doi:10.1016/j.cell.2019.07.008

Pulikkan, J., Mazumder, A., Grace, T. (2019) 'Role of the Gut Microbiome in Autism Spectrum Disorders.' In: Guest P. (ed.) *Reviews on Biomarker Studies in Psychiatric and Neurodegenerative Disorders. Advances in Experimental Medicine and Biology*, vol. 1118. Springer, Cham. https://doi.org/10.1007/978-3-030-05542-4_13

Chu, C. *et al.* (2019) 'The microbiota regulate neuronal function and fear extinction learning.' *Nature* 574, 543–8 https://doi.org/10.1038/s41586-019-1644-y

Callaghan, B. L. *et al.* (2020) 'Mind and gut: Associations between mood and gastrointestinal distress in children exposed to adversity.' *Dev Psychopathol.* 32(1):309–28. doi:10.1017/S0954579419000087. PMID: 30919798; PMCID: PMC6765443

Clapp, M. *et al.* (2017) 'Gut microbiota's effect on mental health: The gut-brain axis.' *Clin Pract.* 7(4):987. doi:10.4081/cp.2017.987

Vogel, S. C., Brito, N. H. and Callaghan, B. L. (2020) 'Early Life Stress and the Development of the Infant Gut Microbiota: Implications for Mental Health and Neurocognitive Development.' *Current Psychiatry Reports.* 22–61

Elsey, J. W. B., van Ast, V. A., Kindt, M. (2018) 'Human memory reconsolidation: A guiding framework and critical review of the evidence.' *Psychol Bull.* 144(8):797–848. doi:10.1037/bul0000152. Epub 24 May 2018. PMID: 29792441

Brunet, A. *et al.* (2008) 'Effect of post-retrieval propranolol on psychophysiologic responding during subsequent script-driven traumatic imagery in post-traumatic stress disorder.' *J Psychiatr Res.* 42(6):503–6. doi:10.1016/j.jpsychires.2007.05.006. Epub 22 Jun 2007. PMID: 17588604

Brunet, A. *et al.* (2018) 'Reduction of PTSD Symptoms With Pre-Reactivation Propranolol Therapy: A Randomized Controlled Trial.' *Am J Psychiatry.* 175(5):427–33. doi:10.1176/appi.ajp.2017. 17050481. Epub 12 Jan 2018. PMID: 29325446

Thierrée, S. *et al.* (2020) 'Trauma reactivation under propranolol among traumatized Syrian refugee children: preliminary evidence regarding efficacy.' *Eur J Psychotraumatol.* 11(1):1733248. doi:10.1080/2000819 8.2020.1733248. PMID: 32194925; PMCID: PMC7067198

Brunet, A. *et al.* (2019) 'Paris MEM: a study protocol for an effectiveness and efficiency trial on the treatment of traumatic stress in France after the 2015–16 terrorist attacks.' *BMC Psychiatry.* 19(1):351. doi:10.1186/s12888-019-2283-4. PMID: 31703570; PMCID: PMC684 2179

Vaiva, G. *et al.* (2003) 'Immediate treatment with propranolol decreases posttraumatic stress disorder two months after trauma.' *Biol Psychiatry.* 54(9):947–9. doi:10.1016/s0006-3223(03)00412-8

'Can a blood pressure drug help ease the painful memory of an ex?' https://www.bbc.com/news/world-us-canada-51317388

Lonergan, M. *et al.* (2016) 'Reactivating addiction-related memories under propranolol to reduce craving: A pilot randomized controlled trial.' *J Behav Ther Exp Psychiatry.* 50:245–9. doi:10.1016/j. jbtep.2015.09.012. Epub 22 Oct 2015. PMID: 26454715

Chalkia, Anastasia *et al.* (2019) 'Acute but Not Permanent Effects of Propranolol on Fear Memory Expression in Humans.' *Front Hum Neurosci.* https://doi.org/10.3389/fnhum.2019.00051

Roullet, P. *et al.* (2021) 'Traumatic memory reactivation with or without propranolol for PTSD and comorbid MD symptoms: a randomised clinical trial.' *Neuropsychopharmacol.* https://doi.org/10.1038/s41386-021-00984-w

Brunet, A. *et al.* (2008) 'Effect of post-retrieval propranolol on psychophysiologic responding during subsequent script-driven traumatic imagery in post-traumatic stress disorder.' *J Psychiatr Res.* 42(6):503–6. doi:10.1016/j.jpsychires.2007.05.006. Epub 22 Jun 2007. PMID: 17588604

Critchlow, Hannah (2018) *Consciousness: A Ladybird Expert Book* (The Ladybird Expert Series 29), London, UK: Michael Joseph

LeDoux, Joe (2019) *The Deep History of Ourselves: The Four-Billion-Year Story of How We Got Conscious Brains*: Viking

'What's the use of consciousness? How the stab of conscience made us really conscious.' (2016) In *Where's The Action? The Pragmatic Turn in Cognitive Science*, Engel, Andreas K., Friston, Karl and Kragic, Danica (eds) Cambridge, Mass: MIT Press. doi:10.7551/mitpress/9780262034326.003.0012

Marshall, P. R. *et al.* (2020) 'Dynamic regulation of Z-DNA in the mouse prefrontal cortex by the RNA-editing enzyme Adar1 is required for fear extinction.' *Nat Neurosci* 23, 718–729 https://doi.org/10.1038/s41593-020-0627-5

Carhart-Harris, R. L. *et al.* (2016) 'Neural correlates of the LSD experience revealed by multimodal neuroimaging.' *Proc Natl Acad Sci USA.*113(17):4853–8. doi:10.1073/pnas 1518377113

Kringelbach, Morten L. *et al.* (2020) 'Dynamic coupling of whole-brain neuronal and neurotransmitter systems.' *Proceedings of the National Academy of Sciences*, 117 (17) 9566–76; doi:10.1073/pnas.1921475117

Atasoy, S., Donnelly, I. and Pearson, J. (2016) 'Human brain networks function in connectome-specific harmonic waves.' *Nat Commun* 7, 10340 https://doi.org/10.1038/ncomms10340

Atasoy, S. *et al.* (2017) 'Connectome-harmonic decomposition of human brain activity reveals dynamical repertoire re-organization under LSD.' *Sci Rep* 7, 17661 https://doi.org/10.1038/s41598-017-17546-0

https://qualiacomputing.com/2017/06/18/connectome-specific-harmonic-waves-on-lsd

Luppi, A. I. *et al.* 'Connectome Harmonic Decomposition of Human Brain Dynamics Reveals a Landscape of Consciousness.' Pre-print. Posted 10 August 2020. doi:https://doi.org/10.1101/2020.08.10.244459

Forstmann, M. *et al.* (2020) 'Transformative experience and social connectedness mediate the mood-enhancing effects of psychedelic use in naturalistic settings.' *PNAS* 20 117 (5) 2338–46 https://doi.org/10.1073/pnas.1918477117

Carhart-Harris, Robin L. *et al.* (2016) 'Neural correlates of the LSD experience revealed by multimodal neuroimaging.' *PNAS* 113 (17) 4853–8 https://doi.org/10.1073/pnas.1518377113

Nutt, D., Erritzoe, D. and Carhart-Harris, R. (2020) 'Psychedelic Psychiatry's Brave New World.' *CELL*, vol. 181, 24–28, ISSN: 0092-8674

Revenga, Mario de la Fuente *et al.* (2021) 'Prolonged epigenetic and synaptic plasticity alterations following single exposure to a psychedelic in mice.' bioRxiv.02.24.432725; doi:https://doi.org/10.1101/2021.02.24.432725

Healing as Collective Intelligence

Sacks, Jonathan (2020) *Morality: Restoring the Common Good in Divided Times.* London, UK: Hodder & Stoughton

Van Steenbergen, H. *et al.* (2021) 'How positive affect buffers stress responses.' *Current Opinion in Behavioral Sciences* 39: 153160

Ioannidis, K. *et al.* (2020) 'The complex neurobiology of resilient functioning after childhood maltreatment.' *BMC Medicine* 18: e32

Van Harmelen, A. L. *et al.* (2016) 'Friendships and family support reduce subsequent depressive symptoms in at-risk adolescents.' *PLoS One* 11(5)

Van Harmelen A. L. *et al.* (2017) 'Adolescent friendships predict later resilient functioning across psychosocial domains in a healthy community cohort.' *Psychological Medicine* 47(13): 2312–22

Does laughing alter your brain chemistry? https://www.thenakedscientists.com/podcasts/naked-neuroscience/neuroscience-nuggets-2013

Scott, Sophie, 'Why We Laugh.' TEDx talk, March 2015 https://www.ted.com/talks/sophie_scott_why_we_laugh?language=en

Catron, Mandy Len (2015) 'To Fall in Love With Anyone, Do This.' *New York Times*, 9 January. Retrieved 11 January 2015.

Eger, Edith (2020) *The Gift: 12 Lessons to Save Your Life*. Rider.

Eger, Edith (2018) *The Choice: A True Story of Hope*. Rider.

Mandela, Nelson (1995) *Long Walk To Freedom: The Autobiography of Nelson Mandela* (new edition). Abacus.

Ubuntu: (I am because we are) Philosophy: A Road to 'individualism' to global solidarity, https://www.academia.edu/45015997/Ubuntu_I_am_because_we_are_Philosophy_A_Road_to_individualism_to_global_solidarity

Arai, Tatsushi and Niyonzima, Jean Bosco (2019) 'Learning Together to Heal: Toward an Integrated Practice of Transpersonal Psychology, Experiential Learning, and Neuroscience for Collective Healing.' *Peace and Conflict Studies* vol. 26, no. 2, article 4. Available at: https://nsuworks.nova.edu/pcs/vol26/iss2/4

Carstarphen, N. (2004) 'Making the other human: The role of personal stories to bridge deep differences.' In Slavik, H. (ed.), *International Communication and Diplomacy* 177–96. Malta and Geneva: Diplo Foundation

Volkan, V. (2004) *Blind Trust: Large Groups and Their Leaders in Times of Crisis and Terror.* Charlottesville, VA: Pitchstone Publishing

Gillson, S. L., Ross, D. A. (2019) 'From Generation to Generation: Rethinking "Soul Wounds" and Historical Trauma.' *Biol Psychiatry*. 86(7):e19–e20. doi:10.1016/j.biopsych.2019.07.033. PMID: 31521209; PMCID: PMC7557912

Bruce Springsteen speaking to Barack Obama on the podcast *Renegades: Born in the USA*: 'Our unlikely friendship'. 39 minutes in. https://open.spotify.com/show/42xagXCUDsFO6aolcHoTlv

10. Us and AI: Transhumanism and Merging Minds

'Meet the caring Robot trio!' Interview with Professor Goldie Nejat, University of Toronto, 17 November 2014: https://www.thenakedscientists.com/articles/interviews/meet-caring-robot-trio

'Boris Johnson pledges £250m for NHS artificial intelligence.' https://www.theguardian.com/society/2019/aug/08/boris-johnson-pledges-250m-for-nhs-artificial-intelligence

'Can Artificial Intelligence Help See Cancer in New, and Better, Ways?' https://www.cancer.gov/news-events/cancer-currents-blog/2022/artificial-intelligence-cancer-imaging 22 March 2022, by NCI staff

'A robot wrote this entire article. Are you scared yet, human?' GPT-3. https://www.theguardian.com/commentisfree/2020/sep/08/robot-wrote-this-article-gpt-3

https://projects.iq.harvard.edu/rak/event/turing-test-poetry-fest, Turing Test Poetry Fest, Monday, October 25, 2021

Rockmore, Dan (2020) 'What Happens When Machines Learn to

Write Poetry.' *New Yorker*, 7 January. https://www.newyorker.com/culture/annals-of-inquiry/the-mechanical-muse

'Deep Blue vs Kasparov: How a computer beat best chess player in the world.' BBC News: https://www.youtube.com/watch?v=KF6sLCeBjos

Critchlow, Hannah (2018) *Consciousness: A Ladybird Expert Book* (The Ladybird Expert Series 29), London, UK: Michael Joseph

AI as good as humans at keeping us alive? *Lancet* study says yes . . . https://www.sciencemediacentre.org/expert-reaction-to-a-study-looking-at-the-effectiveness-of-ai-at-diagnosing-disease-compared-to-health-professionals

Can AI predict the future? https://www.newscientist.com/article/mg24332500-800-predicting-the-future-is-now-possible-with-powerful-new-ai-simulations

Artificial intelligence can predict premature death, study finds: https://www.sciencedaily.com/releases/2019/03/190327142032.htm

https://www.theguardian.com/books/2018/jun/15/rise-of-the-machines-has-technology-evolved-beyond-our-control-

AI in NHS: some good examples with academic link, https://www.bbc.co.uk/news/health-49270325?fbclid=IwAR2NnvJjC8QIvBhw1F6Il5lmTG3vKv3UbQ5rHQbK-5ICb-7hvhqrySOAM2M

'AI Can Detect Signals for Mental Health Assessment.' https://scitechdaily.com/ai-can-detect-signals-for-mental-health-assessment

Von Radowitz, John (2017) 'Intelligent machines will replace teachers within 10 years, leading public school headteacher predicts.' *Independent*, 11 September https://www.independent.co.uk/tech/intelligent-machines-replace-teachers-classroom-10-years-ai-robots-sir-anthony-sheldon-wellington-college-a7939931.html

Andrews, Mark (2022) 'The delivery drone revolution that is sweeping the country.' *Express & Star*, 8 January https://www.expressandstar.com/news/business/2022/01/08/watch-out-the-drones-are-coming

https://www.independent.co.uk/life-style/gadgets-and-tech/news/artificial-intelligence-human-sleep-ai-los-alamos-neural-network-a9554271.html

'World first as artificial neurons developed to cure chronic diseases.' Press release, 3 December 2019: https://www.bath.ac.uk/

announcements/world-first-as-artificial-neurons-developed-to-cure-chronic-diseases

Tsvetkova, Milena *et al.* (2017) 'Even Good Bots Fight: The case of Wikipedia.' *PLoS One* https://doi.org/10.1371/journal.pone.0171774

'Meet the caring Robot trio!' Interview with Professor Goldie Nejat, University of Toronto, 17 November 2014: https://www.thenakedscientists.com/articles/interviews/meet-caring-robot-trio

Stephen Hawking warns artificial intelligence could end mankind, December 2014 https://www.bbc.co.uk/news/technology-30290540

Elon Musk claims AI will overtake humans 'in less than five years', 27 July 2020 https://www.independent.co.uk/tech/elon-musk-artificial-intelligence-ai-singularity-a9640196.html

Kasparov, Garry (2017) *Deep Thinking: Where Machine Intelligence Ends and Human Creativity Begins.* London, UK: John Murray

What Can We Do Already?

Lozano, A. M. *et al.* (2019) 'Deep brain stimulation: current challenges and future directions.' *Nat Rev Neurol.* 15(3):148–60. doi:10.1038/s41582-018-0128-2. PMID: 30683913; PMCID: PMC6397644

Krauss, J. K. *et al.* (2021) 'Technology of deep brain stimulation: current status and future directions.' *Nat Rev Neurol* 17, 75–87 https://doi.org/10.1038/s41582-020-00426-z

Graat, I., Figee, M., Denys, D. (2017) 'The application of deep brain stimulation in the treatment of psychiatric disorders.' *Int Rev Psychiatry.* 29(2):178–90. doi:10.1080/09540261.2017.1282439. Epub 10 Feb 2017. PMID: 28523977

Seo, Dongjin *et al.* (2016) 'Wireless Recording in the Peripheral Nervous System with Ultrasonic Neural Dust.' *Neuron*, vol. 91, issue 3, 529–39 doi:https://doi.org/10.1016/j.neuron.2016.06.034,

Patch, K. (2021) 'Neural dust swept up in latest leap for bioelectronic medicine.' *Nat Biotechnol* 39, 255–6 https://doi.org/10.1038/s41587-021-00856-0

Strollo, Patrick J. *et al.* (2014) 'Upper-Airway Stimulation for Obstructive Sleep Apnea.' *New England Journal of Medicine.* 370 (2): 13949. doi:10.1056/NEJMoa1308659. ISSN 0028-4793

Osorio, I. *et al.* (2001) 'An introduction to contingent (closed-loop) brain electrical stimulation for seizure blockage, to ultra-short-term clinical trials, and to multidimensional statistical analysis of therapeutic efficacy.' *Journal of Clinical Neurophysiology.* 18 (6): 533–44. doi:10.1097/00004691-200111000-00003. ISSN 0736-0258. PMID 11779966

Chow, Alan Y. (2004) 'The Artificial Silicon Retina Microchip for the Treatment of VisionLoss From Retinitis Pigmentosa.' *Archives of Ophthalmology.* 122 (4): 460–9. doi:10.1001/archopht.122.4.460. ISSN 0003-9950. PMID:15078662

Hochberg, Leigh R. *et al.* (2006) 'Neuronal ensemble control of prosthetic devices by a human with tetraplegia.' *Nature.* 442 (7099): 164–71. Bibcode:2006Natur.442.164H. doi:10.1038/nature04970. ISSN 1476-4687. PMID 16838014

Keene, S. T. *et al.* (2020) 'A biohybrid synapse with neurotransmitter-mediated plasticity.' *Nat Mater.* 19, 969–73 https://doi.org/10.1038/s41563-020-0703-y

Lu, Q. *et al.* (2020) 'Biological receptor-inspired flexible artificial synapse based on ionic dynamics.' *Microsyst Nanoeng* 6, 84 https://doi.org/10.1038/s41378-020-00189-z

Berger, T. W. *et al.* (2012) 'A hippocampal cognitive prosthesis: multi-input, multi-output nonlinear modeling and VLSI implementation.' *IEEE Trans Neural Syst Rehabil Eng.* 20(2):198–211. doi:10.1109/TNSRE.2012.2189133. PMID: 22438335; PMCID: PMC3395724

Hampson, Robert E. *et al.* (2018) 'Developing a hippocampal neural prosthetic to facilitate human memory encoding and recall.' *J. Neural Eng.* 15 036014

Hampson, R. E. *et al.* (2018) 'A hippocampal neural prosthetic for restoration of human memory function.' *Journal of Neural Engineering,* 15, 036014, doi:10.1088/1741-2552/aaaed7

Liu, J., *et al.* (2015) 'Syringe-injectable electronics.' *Nature Nanotech* 10, 629–36 https://doi.org/10.1038/nnano.2015.115

Fu, T. M. *et al.* (2016) 'Stable long-term chronic brain mapping at the single-neuron level.' *Nat Methods* 13, 875–82 https://doi.org/10.1038/nmeth.3969

Hong G. *et al.* (2018) 'Mesh electronics: a new paradigm for tissue-like brain probes.' *Curr Opin Neurobiol.* 50:33–41. doi:10.1016/j.

conb.2017.11.007. Epub 1 Dec 2017. PMID: 29202327; PMCID: PMC5984112

Zhou, Tao *et al.* (2017) 'Syringe-injectable mesh electronics integrate seamlessly with minimal chronic immune response in the brain.' *PNAS* https://doi.org/10.1073/pnas.1705509114

Sharon, A. *et al.* (2021) 'Ultrastructural Analysis of Neuroimplant-Parenchyma Interfaces Uncover Remarkable Neuroregeneration Along-With Barriers That Limit the Implant Electrophysiological Functions.' *Front Neurosci.* 15:764448. doi:10.3389/fnins.2021.764448. PMID: 34880722; PMCID: PMC8645653

Neuralink: https://neuralink.com

Elon Musk claims Neuralink's brain implants will 'save' memories like photos and help paraplegics walk again. Here's a reality check: https://fortune.com/2022/02/22/elon-musk-neuralink-brain-implant-claims February 22, 2022 5:37 PM GMT

'Don't be brainwashed – Elon Musk's "bionic pig" is just a publicity stunt.' https://www.theguardian.com/commentisfree/2020/sep/01/elon-musk-bionic-pig-publicity-stunt-innovations

'Is Elon Musk over-hyping his brain-hacking Neuralink tech?' https://www.bbc.com/news/technology-53987919

'Elon Musk talks Twitter, Tesla and how his brain works – live at TED2022.' https://www.ted.com/talks/elon_musk_elon_musk_talks_twitter_tesla_and_how_his_brain_works_live_at_ted2022

'Neuralink: Elon Musk unveils pig with chip in its brain.' 29 August 2020

https://www.bbc.co.uk/news/world-us-canada-53956683

Professor Peter Robinson research: https://www.cl.cam.ac.uk/research/rainbow/emotions

Hire View: https://www.hirevue.com/demo/full-platform-em?utm_source=google&utm_medium=cpc&utm_campaign=G___Brand_-_Exact_EMEA_UK_&_Ireland&utm_term=hirevue&gclsrc=aw.ds&gclid=CjwKCAjw9-KTBhBcEiwAr19ig5wcHBEyZjLY1IzexDdd HHTHCiNSKCocnr3bWQ9IwPqjm69fe2dO3RoCZgkQAvD_BwE

Oxygen Forensics: https://www.oxygen-forensic.com/en

Cogito: https://cogitocorp.com

REFERENCES

AI Now Annual Report 2019: https://ainowinstitute.org/AI_Now_2019_Report.pdf

'Emotion-detecting tech should be restricted by law – AI Now.' 12 December 2019 https://www.bbc.com/news/technology-50761116

'Paraplegic in robotic suit kicks off World Cup' 12 June 2014 https://www.bbc.co.uk/news/science-environment-27812218

Carmena, J. M. *et al.* (2003) 'Learning to control a brain-machine interface for reaching and grasping by primates.' *PLOS Biology*, 1 (2): 193–208, doi:10.1371/journal.pbio.0000042, PMC 261882, PMID:14624244

Lebedev, M. A. *et al.* (2005) 'Cortical ensemble adaptation to represent actuators controlled by a brain machine interface.' *J. Neurosci.* 25 (19): 4681–93, doi:10.1523/jneurosci.4088-04.2005, PMC 6724781, PMID:15888644

Nicolelis, Miguel Ângelo Laporta (2003) 'Brain-machine interfaces to restore motor function and probe neural circuits.' *Nat Rev Neurosci*, 4 (5): 417–22, doi:10.1038/nrn1105, PMID:12728268, S2CID:796658

Pais-Vieira, M. *et al.* (2013) 'Brain-to-Brain Interface for Real-Time Sharing of Sensorimotor Information.' *Scientific Reports* 3, 1319

O'Doherty, J. E. *et al.* (2011) 'Active tactile exploration using a brain-machine-brain interface.' *Nature.* 479(7372):228–31. doi:10.1038/nature10489. PMID: 21976021; PMCID: PMC3236080

Pais-Vieira, M. *et al.* (2015) 'Building an organic computing device with multiple interconnected brains.' *Sci Rep.* 5:11869. doi:10.1038/srep11869. Erratum in: *Sci Rep.* 2015;5:14937. PMID: 26158615; PMCID: PMC4497302

Ramakrishnan, A. *et al.* (2015) 'Computing Arm Movements with a Monkey Brainet.' *Sci Rep* 5, 10767 https://doi.org/10.1038/srep10767

Renton, Angela I., Mattingley, Jason B. and Painter, David R. (2019) 'Optimising non-invasive brain-computer interface systems for free communication between naïve human participants.' *Scientific Reports*, 9 (1) 18705, 18705. doi:10.1038/s41598-019-55166-y

Jiang, L. *et al.* (2019) 'BrainNet: A Multi-Person Brain-to-Brain Interface for Direct Collaboration Between Brains.' *Sci Rep* 9, 6115 https://doi.org/10.1038/s41598-019-41895-7

Grau, Carles *et al.* (2014) 'Conscious Brain-to-Brain Communication in Humans Using Non-Invasive Technologies.' *PLoS One* 9 (8): e105225 doi:10.1371/journal.pone.0105225

Rao R. P. *et al.* (2014) 'A direct brain-to-brain interface in humans.' *PLoS One* 9:e111332. 10.1371/journal.pone.0111332

Grau, Carles *et al.* (2014) 'Conscious brain-to-brain communication in humans using non-invasive technologies.' *PLoS One.* 9(8):e105225. doi:10.1371/journal.pone.0105225. eCollection 2014

Stocco, A. *et al.* (2015). 'Playing 20 questions with the mind: collaborative problem solving by humans using a brain-to-brain interface.' *PLoS One* 10:e0137303. 10.1371/journal.pone.0137303

O'Doherty, J. E. *et al.* (2011) 'Active tactile exploration using a brain–machine–brain interface.' *Nature* 479: 228–31

Hsiao, M-C., Song, D. and Berger, T. W. (2013) 'Nonlinear dynamical model based control of in vitro hippocampal output.' *Frontiers in Neural Circuits* 7, 20:1–14.

Sun, Chen *et al.* 'Hippocampal neurons represent events as transferable units of experience.' *Nature Neuroscience* doi:10.1038/s41593-020-0614-x. https://neurosciencenews.com/memory-situation-neurons-16085

Deadwyler, S. A. and Hampson, R.E. (2006) 'Temporal coupling between subicular and hippocampal neurons underlies retention of trial-specific events.' *Behav Brain Res.* 174, 272–80. doi:10.1016/j.bbr.2006.05.038

Deadwyler, S. A., Goonawardena, A. V. and Hampson, R. E. (2007) 'Short-term memory is modulated by the spontaneous release of endocannabinoids: evidence from hippocampal population codes.' *Behav Pharm* 18, 571–80. doi:10.1097/FBP.0b013e3282ee2adb

Deadwyler, S. A. *et al.* (2013) 'Donor/recipient enhancement of memory in rat hippocampus.' *Front Syst Neurosci* 7: 120

Qiao, Z. *et al.* (2018) 'ASIC Implementation of a Nonlinear Dynamical Model for Hippocampal Prosthesis.' *Neural Comput.* 30(9):2472–99. doi:10.1162/neco_a_01107. Epub 27 Jun 2018. PMID: 29949460

Li, W. X. *et al.* (2013) 'Real-time prediction of neuronal population spiking activity using FPGA.' *IEEE Trans Biomed Circuits Syst.* 7(4):489–98. doi:10.1109/TBCAS.2012.2228261. PMID: 23893208

Berger, T. W. *et al.* (2005) 'Restoring lost cognitive function.' *IEEE*

Engineering in Medicine and Biology Magazine vol. 24, no. 5, 30–44 doi:10.1109/MEMB.2005.1511498

Entering an Ethical Quagmire

Khan, S. and Aziz, T. (2019) 'Transcending the brain: is there a cost to hacking the nervous system?' *Brain Commun.*1(1):fcz015. doi:10.1093/braincomms/fcz015. eCollection 2019. PMID: 32954260

Perbal, B. (2015) 'Ethical considerations of BBI: 'Knock once for yes, twice for no".' *J Cell Commun Signal.* 9(1):15–8. doi:10.1007/s12079-015-0273-y. Epub 26 Feb 2015. PMID: 25711904

Vansteensel, M. J. *et al.* (2016) 'Fully implanted brain–computer interface in a locked-in patient with ALS.' *N Engl J Med* 375: 2060–6

Trimper, J. B., Wolpe, P. R. and Rommelfanger, K. S. (2014) 'When "I" becomes "We": ethical implications of emerging brain-to-brain interfacing technologies.' *Front Neuroeng* 7 4: 1–4

Moses, D. A. *et al.* (2019) 'Real-time decoding of question-and-answer speech dialogue using human cortical activity.' *Nat Commun* 10, 3096 https://doi.org/10.1038/s41467-019-10994-4

Wouters, Niels and Vetere, Frank, University of Melbourne, 'Holding a Black Mirror Up to Artificial Intelligence.' https://pursuit.unimelb.edu.au/articles/holding-a-black-mirror-up-to-artificial-intelligence

(2019) iHuman: a futuristic vision for the human experience.' *The Lancet* vol. 394, issue 10203, P979.

Royal Society iHuman perspective: Neural interfaces, https://royalsociety.org/-/media/policy/projects/ihuman/report-neural-interfaces.pdf, 10 September 2019, https://royalsociety.org/topics-policy/projects/ihuman-perspective

Melding Minds to Dream Much Bigger

Cell Press (2020) 'How "swapping bodies" with a friend changes our sense of self.' *Science Daily.* www.sciencedaily.com/releases/2020/08/200826110322.htm

Tacikowski, Pawel, Weijs, Marieke L., Ehrsson, Henrik H. (2020) 'Perception of Our Own Body Influences Self-Concept and Self-Incoherence Impairs Episodic Memory.' *iScience* 101429 doi:10.1016/j.isci.2020.101429

Roche, Richard 'Emotional Contagion, 3 person closed loop system' https://vimeo.com/237065016

The Communal Kiss, 'The Communal Kiss': https://vimeo.com/332483454

Empathy Ecologies: https://vimeo.com/422895498

'Tribute to Jose Delgado, Legendary and Slightly Scary Pioneer of Mind Control' https://blogs.scientificamerican.com/cross-check/tribute-to-jose-delgado-legendary-and-slightly-scary-pioneer-of-mind-control

Epilogue

Bostrom, Nick (2006) 'What is a Singleton?' *Linguistic and Philosophical Investigations*, Vol. 5, No. 2, pp. 48–54

Critchlow, H.M. (2007) 'Investigating the Role of Dendritic Spine Plasticity in Schizophrenia', doctoral thesis, University of Cambridge.

Fone K.C., Porkess M.V. (2008) 'Behavioural and neurochemical effects of post-weaning social isolation in rodents-relevance to developmental neuropsychiatric disorderbs.' *Neurosci Biobehav Rev*; 32(6): 1087-102. doi: 10.1016/j.neubiorev.2008.03.003. PMID: 18423591.

Kini, Naren 'My Neighbours Corn', Awakin, https://www.awakin.org/v2/read/view.php?tid=2395

Ratner, Paul (2008) The "singleton hypothesis" predicts the future of humanity, *Big Think* https://bigthink.com/the-present/singleton-hypothesis-future-humanity/

Reser, David, Simmons, Margaret, Johns, Esther, Ghaly, Andrew, Quayle, Michelle, Dordevic, Aimee L., Tare, Marianne, McArdle, Adelle, Willems, Julie and Yunkaporta, Tyson 2021, Australian Aboriginal techniques for memorization: translation into a medical and allied health education setting, PLoS One, vol. 16.

Sand Talk: How Indigenous Thinking Can Save the World, by Tyson Yunkaporta, HarperOne; Illustrated edition (12 May 2020) ISBN-13 : 978-0062975645

Waring, Timothy M., Wood, Zachary T. (1952) 'Long-term gene–culture coevolution and the human evolutionary transition.' *Proceedings of the Royal Society B: Biological Sciences*, 2021; 288: 20210538 DOI: "http://dx.doi.org/10.1098/rspb.2021.0538" 10.1098/rspb.2021.0538

Index